THE
EXHAUSTED
BRAIN

THE
EXHAUSTED
BRAIN

The Origin of Our Mental Energy
and Why It Dwindles

Michael Nehls, MD, PhD

Translated by Andy Jones Berasaluce

Skyhorse Publishing

Skyhorse Publishing books may be purchased in bulk at special discounts for sales promotion, corporate gifts, fund-raising, or educational purposes. Special editions can also be created to specifications. For details, contact the Special Sales Department, Skyhorse Publishing, 307 West 36th Street, 11th Floor, New York, NY 10018 or info@skyhorsepublishing.com.

Skyhorse® and Skyhorse Publishing® are registered trademarks of Skyhorse Publishing, Inc.®, a Delaware corporation.

Visit our website at www.skyhorsepublishing.com.

Please follow our publisher Tony Lyons on Instagram @tonylyonsisuncertain.

10 9 8 7 6 5 4 3 2 1

Library of Congress Cataloging-in-Publication Data is available on file.

Print ISBN: 978-1-5107-8302-7
eBook ISBN: 978-1-5107-8303-4

Jacket design by David Ter-Avanesyan
Jacket image by Shutterstock

Printed in the United States of America

For Sabine, Sebastian, Sarah, and Nadja—my four-leaf clover

CONTENTS

INTRODUCTION

All that we are is the result of what we have thought.
—Buddha

All Life Is Problem Solving is the title of a book by the Austrian-British philosopher Sir Karl R. Popper (1902–1994).[1] This sounds plausible, because after all, our actions, as well as our inactions, are always preceded by a decision—in other words, the answer, whether conscious or unconscious, to the question of what to do next. Nothing we do happens purely by chance. This also applies to the environmental destruction we cause, the pollution of the oceans, the contamination of the air we breathe, and the ever-accelerating extinction of species, to which a large part of humanity will sooner or later fall victim. Our downfall would therefore also be a consequence of human decisions.

All Life Is Problem Creating would therefore have been an alternative title for Popper's book. But why do we behave so contrary to human nature? After all, we should all pursue the existential goal of paving the way for our descendants into a healthy future, instead of destroying their livelihood. It is often claimed that these catastrophic developments are the result of the work of greedy business leaders and corrupt politicians—as if they were forcing their self-destructive lifestyle on humanity. The capitalist system is also often cited as the cause—as if other economic systems were actual alternatives. But fundamentally, such explanations only serve to absolve us of personal responsibility. After all, it is we, the vast majority of "normal" people, who decide what we buy, what we eat, who we elect, and to whom we grant political power over us.

When I use the word "we" or write about "us," I mean at least a decisive majority in relation to the respective behavior or circumstance. Incidentally, I myself am no exception to this "we/us," as you can see from many of the anecdotes from my life that precede the chapters. But I persistently try to understand the reasons why I, too, repeatedly made wrong decisions, if only to be able to make better ones in the future—very likely, this is the only reason this book exists. It is my attempt to find a fundamental answer to the question of why necessary lifestyle changes are so difficult for me, and therefore presumably also for us, meaning a large part of humanity, and why we repeatedly abandon good intentions or simply make the right decisions too rarely.

A central question that needs to be answered is therefore: "Would you be willing to change your current way of life if it turned out that it would endanger your own life and that of your children in the long run (not immediately, as the answer would be simple)?" For many people, the answer to this question would be a no (or more often a yes that would never be seriously implemented). The proof that this is true for a majority of people can be seen in the medical field. After all, not only is the destruction of our environment and thus our livelihood a consequence of human behavior, but so are almost all so-called diseases of civilization such as diabetes, arteriosclerosis with stroke or heart attack, Alzheimer's, and many types of cancer. Yet, although most people suffer from these preventable diseases and will ultimately die because there are no cures for them, few are willing to prevent the onset of these diseases by making timely changes to their lifestyle. Obviously, even chronic illness and premature death are not incentive enough to give up cherished but ultimately harmful habits. With such a self-destructive attitude, the expectation that there will be no change in lifestyle when it comes to such general and abstract things as ecological balance of fair future opportunities for all children of this Earth is not surprising. Why is that? Do most people have their own personal reasons for this, so that it's just a coincidence that almost everyone behaves uniformly, or is there a more fundamental explanation for it?

Change requires alternative thinking and knowledge-based action, yet more and more people have almost a fear of the new and thus also of change, even when it is urgently needed. Perhaps there is also a lack of creativity and imagination to envision an alternative way of life. And even when the necessity for change is recognized as indispensable, there is then a lack of willpower in its implementation. Overcoming fear and daring to try something new requires mental energy, which obviously far too many people cannot muster. What kind of energy might this be? Along the same lines, I was once asked whether our brain can actually get tired or whether our concentration decreases when we are mentally active. At first, I was a bit perplexed by the triviality of this question; after all, the questioner, if not an extraterrestrial, would have to experience this himself every day. But I simply replied politely that mental work requires energy, which is the reason why we also need to rest and recover again and again.

But something about the question gnawed at me, because after all, it is true that this answer does not apply to the entire brain. For example, one area of our brain regulates breathing—without interruption and without tiring. If this part of the brain required the same mental energy that our thinking and acting need, there would always be the risk that we would suffocate from exhaustion. So, the somewhat more specific question is, why does our decision-making center, which is located directly behind our "thinker's forehead" in the frontal lobe, get tired? What kind of energy does it consume that other areas of the brain that are constantly active apparently do not?

In recent decades, psychological research has discovered that the mental energy available to our frontal lobe is, in fact, finite. This limit affects our thinking, decision-making, and actions. When our mental energy store is empty, this doesn't just have a quantitative effect, as with muscle that, when it lacks energy, can simply move less weight. It also has qualitative effects: One is less willing to think outside the box. Instead, one tends toward stereotypical thinking, is trapped in routines, and thereby loses an ability that actually distinguishes us humans, namely being able to meaningfully adapt to changing situations. Creativity, imagination, and willpower are diminished, and even self-esteem suffers. Scientifically, this state is referred to as "ego depletion," which means nothing other than "mental exhaustion." For the contribution of these findings from psychological research, particularly to economics regarding "judgments and decisions under uncertainty," the Israeli American psychologist Daniel Kahneman and the American economist Vernon L. Smith shared the Nobel Prize in Economics in 2002. However, the two left the question unanswered as to what kind of mental energy our frontal lobe needs for decision-making, but also for creativity, long-term planning, and willpower to achieve goals. Even after extensive research, almost twenty years later (as of January 2022), this crucial question remains, as does the question of where in our brain this mental energy is stored.

To identify this previously unknown "frontal lobe battery," as I would like to call the storage location of this mental energy, I used all previously known functions of this energy as a kind of compass—and I actually found it! There is a place in our brain that fulfills all of the functional criteria. It turns out that this type of mental energy has a physical presence. Through the discovery of the so-called frontal lobe battery, presented here for the first time, we not only know the storage location of our mental energy, which our frontal lobe needs, but we also know its nature. We can now explain how the frontal lobe battery "discharges" through thinking and what happens when it "recharges" during sleep. Last but not least, we now also know what limits its storage capacity, and thus why our ego can become depleted.

Countless studies show that we have the genetic potential to increase the frontal lobe battery's storage capacity throughout our lives. This insight is of far-reaching importance, both for us as individuals and for humanity as a whole. After all, our frontal lobe would have more mental energy if it had a stronger frontal lobe battery. This would make us mentally more flexible and help us to make better life decisions for ourselves more often and implement them with more self-confidence and willpower. As it turns out, the activation of this "capacity increase program" is based on an evolutionary logic. Under the living conditions of a very long Paleolithic phase, the brain of *Homo sapiens sapiens* (doubly so because he knows that he knows, i.e. is aware of his thinking), that is, the wise and clever human, developed its highest capacity. Through selection pressure, there was a genetic adaptation to these mind-promoting living conditions. Thus, these became necessities and therefore conditions that must still be met today so that our frontal lobe battery can develop optimally and continue to increase its capacity throughout life. These conditions, in their totality, define a species-appropriate life. To live "species-appropriately," it is entirely sufficient to know and satisfy these natural needs of our brain, without having to live as in the Paleolithic. If we are ready for this, then we can with very simple measures do the following:

- support the natural development of our children so that they can fully develop their mental potential,
- increase creativity, self-confidence, willpower, and perseverance in all phases of life,
- improve memory,
- increase emotional and social intelligence,
- promote rational thinking,
- strengthen psychological resilience,
- increase the natural interest in new experiences and the joy of life,
- perceive things more consciously,
- and, last but not least, protect ourselves from burnout, depression, and even Alzheimer's dementia in a sustainable way.

That's the good news. But there's also bad news. Because our modern way of life deviates significantly from what our brain needs to develop and maintain these functions in many areas, the frontal lobe battery no longer reaches its genetically possible capacity even in childhood. As a result of our now almost completely unnatural way of life, its natural potential to continue increasing its capacity throughout life is also not being utilized. Rather, the frontal lobe battery even continuously loses storage capacity in the "normal" or average adult. A frontal lobe battery's storage capacity, underdeveloped from the outset, that continues to degrade throughout life causes chronic ego depletion, chronic weakening of frontal lobe functions, which affects a large part of humanity. This is highly problematic and explains, among other things:

- the increase in anxiety symptoms and depression, even in children,
- the decline in emotional and social intelligence and the ability to empathize in society at large,
- the increase in collective narcissism with a tendency toward stereotypical thinking and acting,
- and the deeply disturbing fact that we continue as before despite the impending collapse of our livelihood.

The "pandemic of frontal lobe impairment" as a result of a chronic reduction in frontal lobe battery capacity thus provides a neurobiological explanation for many dramatic developments that can be observed worldwide. This unfortunately includes, among other things, the obvious inability of humanity to stop the destruction of its limited habitat. It also explains why so many people fail to change their way of life, even when they know that this would be vital for their own health and their children's future. A significant portion of humanity is thus caught in a vicious circle of frontal lobe impairment, which can probably only be broken by applying the knowledge presented here. If this is not achieved, the self-destructive process on an individual and global level is very likely to be unstoppable.

CHAPTER 1

THE HUMAN BRAIN'S EXECUTIVE CENTER

The Brain—is wider than the Sky—
For—put them side by side—
The one the other will contain
With ease—and you—beside—

The Brain is deeper than the sea—
For—hold them—Blue to Blue—
The one the other will absorb—
As sponges—Buckets—do—

The Brain is just the weight of God—
For—Heft them—Pound for Pound—
And they will differ—if they do—
As syllable from Sound—
—Emily Dickinson (1830–1886)

The Power of Human Imagination

As a young boy, I was fascinated by Robin Hood. I loved crafting bows and arrows and then going duck hunting in the nearby woods. After all, my parents enjoyed eating poultry, and I wanted to do my part as a little Robin to contribute to the family's food supply. To improve my chances of hunting success, I waded through the shallow bed of a small stream where ducks frolicked on the banks. Of course, my equipment was far too primitive to ever be successful in the hunt. But I didn't know that back then, and it didn't matter anyway. All that

1

mattered was the idea and the attempt to implement my idea, coupled with the feeling of being undisturbed in nature, all alone with my thoughts. One question preoccupied me on one of my forays, perhaps due to my constant bad luck in hunting: Does my doing have any meaning when I consider the seemingly infinite vastness of the world around me? *I still remember today where I was trudging along when this question went through my head, and for the first time, an answer came to mind that still seems plausible to me today. It miraculously became clear to me that in nature everything is connected and therefore everything I do, no matter how insignificant it may seem, must have some consequences. Even if I only kicked a stone away on the ground (which I then did), the world was a little different than before. About two decades later, I read James Gleick's book on chaos theory*[1]*—and it was familiar to me from the first to the last page: Every being is like a butterfly, which, as is well known, could trigger a hurricane at any time with just one flap of its wings.*

Creativity is a combination of imagination, vision, ingenuity, and, according to a scientific definition, ultimately the ability to produce original, unusual, flexible, and valuable ideas or behaviors that override an established habit.[2] In fact, we have the ability to understand that everything that is, and everything that happens, could also be completely different or develop differently.

Our brain can even envision things that don't exist at all—and often, they only come to exist due to this act of imagination. This, as the above Emily Dickinson poem so wonderfully expresses, makes our brain a creative maker, larger than the universe, and, perhaps, even a bit larger than God. After all, it could be that our inventive mind also created God in the many cultural and individually conceived variants. So while the Bible says, "In the beginning was the Word" (John 1:1–3), it could also be that it should rather say: "In the beginning was the idea, the thought, the imagination—or simply the human fantasy." Fantasy is the engine of social progress. For Albert Einstein (1879–1955), imagination was even more important than knowledge.[3] This makes perfect sense, because after all, knowledge is limited, but imagination is not. But if, as we know, knowledge means power, how much more powerful must the human imagination be? And if having power entails responsibility, what great responsibility do we have through our imagination? What we do or don't do every day within the framework of our "fantastic" possibilities, both the big and the small decisions that we do or don't make, determines, along with many other factors that are beyond our control, what kind of life we ourselves lead, but also how the lives of many other people go. Even if many things seem to be outside our sphere of influence, we still bear enormous responsibility, even if we are not always aware of it. These thoughts raise many questions. How we answer them will determine the path that humanity takes unlike anything else: Why does the human brain have the potential on the one hand for cultural achievements and on the other hand for alien and self-destructive behaviors? How can it be that extraordinary creativity and social intelligence have enabled *Homo sapiens* to dominate the Earth, only to gradually destroy it so radically that humanity's livelihood and that of its fellow creatures are already seriously threatened and, in some cases, even irrevocably destroyed? How is it that our brain is naturally concerned about the personal well-being of our offspring and yet displays a "devil-may-care" attitude? How can our behavior be characterized by farsighted planning and then again be geared toward the immediate gratification of urges, even if we thereby harm ourselves or other living beings? In fact, many of our brain's decisions repeatedly contradict our own survival instinct—as well as the reproductive instinct, which causes the natural aspiration of every person to always want the best for their offspring, at least. It is therefore of existential importance to find out how the so-called executive center of our brain makes its decisions.

The Frontal Lobe—What Makes Us Human

A pioneer in the discovery and research of the human brain's executive center was—as so often happens with completely new insights throughout the history of medicine—a human tragedy.

On September 13, 1848, Phineas Gage (1823–1860), foreman of a crew of railroad workers in Vermont, was filling a borehole with explosives as usual. Perhaps he wasn't quite focused on this routine task. In any case, he forgot to add insulating sand before he started compacting the powder in the borehole with an iron rod about one meter long and three centimeters thick. Inevitably, the charge exploded, and the rod became a projectile. It pierced his left cheekbone, then the back of his left eye socket, and finally his skull, as the following historical drawing illustrates.[4] Along the way, it destroyed the front part of his cerebrum, the frontal lobe. Curiously, Gage not only remained fully conscious during the accident, he even survived it. According to his doctor, John Martyn Harlow, he was physically recovered after just a few weeks. He had only lost his left eye.

However, a large part of the prefrontal cortex was also destroyed, although not visible from the outside. *Cortex* is Latin for bark and refers here to the cerebral cortex, where nerve cell bodies are predominantly located. The *prefrontal* cortex is the foremost part of the frontal lobe, which I will label as the *frontal lobe* in the following (see Figure 2). Gage's motor skills remained intact after the accident, as they were controlled by the posterior part of the frontal lobe, as did his ability to speak, which was why his doctor initially considered him to have recovered. His perception of his environment and his long-term memory were also unchanged despite the massive brain destruction. Nevertheless, Phineas Gage was not the same after the accident. Previously always known as polite, balanced, and level-headed, he was suddenly changed in character. The destruction of large parts of his prefrontal cortex or his frontal lobe made him unreliable, erratic, and disrespectful—to himself and others. He showed strong mood swings, became a drinker, indulged in aimless pleasures of all kinds, and wasted his money. He didn't seem to care at all about the long-term consequences of his dissolute life.

As tragic as this accident was for him, it was just as revealing for science. The destruction's consequences or the changes in his nature had made it clear what important functions the frontal lobe controls, impacting our life decisions and behavior.[5] It is also astonishing that one can continue to live even after the prefrontal lobe experiences massive destruction—albeit with the loss of some essential qualities that people who think responsibly and act with foresight already possess (this is an important aspect that we will deal with later).

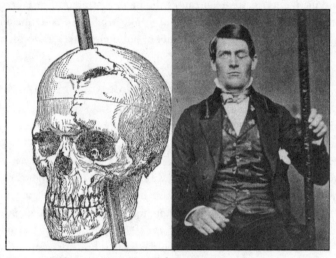

Figure 1: Phineas Gage lived for another twelve years despite extensive destruction of the anterior part of his frontal lobe by an iron rod that pierced his skull in the accident.

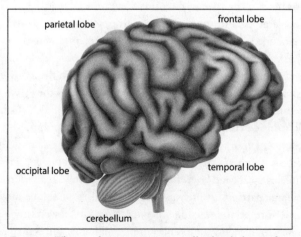

Figure 2: The cerebrum is anatomically divided into four lobes. The frontal lobe or prefrontal cortex is the foremost part of the frontal lobe, located directly behind the forehead.

Many other individual tragedies, triggered by severe brain trauma (accidents), local tumors, or circulatory disorders resulting from arteriosclerosis and stroke, enabled neuroscientists to assign their special functions to the various areas of the frontal lobe.[6] Some of these things impair the ability to analyze and solve problems and to learn from mistakes, others to follow rules or make plans and implement them. For example, it has been discovered that the part directly behind the forehead is responsible for planning, goal-directed behavior, problem-solving, and attention. This region is also the seat of our short-term or working memory. Directly behind it is an area where our "gut feeling" can be localized, but it is also the seat of our social and moral feelings. Damage to this brain area can therefore lead to extreme changes in social behavior. These range from apathy and depression to the extreme opposite: actionism, lack of distance, or even megalomania. All these personality changes, which are associated with different damages to the frontal lobe, are summarized under the general term "frontal lobe syndrome" despite their sometimes very different manifestations. In general, it can be stated that a functioning frontal lobe is a prerequisite for rational intelligence (RQ, derived from IQ), emotional intelligence (EQ), and social intelligence (SQ). If the frontal lobe is damaged or its function is impaired, socially unacceptable actions can often no longer be suppressed.

Interestingly, despite sometimes massive damage to their frontal lobe, those affected usually have no problems with routine actions, as the Gage example already illustrated. For example, shopping for everyday things, preparing meals, and keeping appointments usually proceed completely inconspicuously. If a random conversation partner, whom you do not yet know well, were to suffer from frontal lobe syndrome, you might not even notice it during a superficial conversation.

Frontal Lobe Syndrome or Frontal Lobe Impairment

On closer inspection and in relationships with relatives and generally with the environment, however, it becomes clear that frontal lobe syndrome permanently changes the nature, thinking, and actions of those affected. Thus, it can lead to delusions and paranoid attacks. Therefore, sometimes even relatives seek psychotherapeutic help to better cope with the situation. However, as long as the person with frontal lobe damage does not commit acts of violence, their condition is of secondary importance to society. Fortunately, frontal lobe syndrome is also very rare.

Much more common, however, is a similar change in personality as a result of a lack of mental energy. However, this is not structurally but functionally caused. Since this state of mental exhaustion can affect any of us and influence our thinking and actions, it has much greater social, economic, and political consequences. To distinguish this much more widespread *functional impairment* of the frontal lobe due to a *lack of energy* from frontal lobe syndrome, I will henceforth speak of *frontal lobe weakness* in connection with this functional impairment. This can be acute and temporary, but—as I will show—it can also become chronic.

CHAPTER 2

EGO DEPLETION—ACUTE LOSS OF MENTAL POWER

Go to bed early, tomorrow you have to be friendly.
—Phil Bosmans (1922–2012)

Acute Mental Exhaustion and Its Consequences

When I took over the management of a Munich biotechnology company in the spring of 2000, one of my first actions was to protect my environment from me: I asked my assistant not to put through any more business calls after 5:00 p.m. It's actually in my nature to always take care of things that are pending immediately, if possible. This has the advantage that you don't have to think about it again, which ultimately costs time and energy. But this also includes not avoiding conversations, which has the advantage that I don't have to worry about what the person in question might want. The serious disadvantage, however, is that after a long working day I am exhausted and it is difficult for me not only to adjust to new problems that someone wants to discuss with me, but also to be as friendly and understanding in conversation as is rightfully expected. For the same reason, I don't send any more emails late at night. The next morning, with a clear head, it is simply easier for me to strike the right tone.

Everyone knows what it feels like when the "battery" is empty after a strenuous day. There is even a scientific term for this universally valid neurobiological phenomenon: ego depletion. *Ego* means "I," and *deplere* also comes from Latin and means "to empty." Put together, this means "exhaustion of the self." The reason for this mental exhaustion is that we only have a limited amount of mental energy available to us

7

each day, which is needed by our frontal lobe in particular for all its executive tasks. These include:

- Setting goals and creating a prioritized list when there are multiple goals
- Developing action plans to achieve the primary goal and possible further goals
- Analyzing possible hurdles and difficulties
- Self-motivation and motivating possible partners or team members who may be able to help achieve the goals
- Continuous attention, control, and emotional self-control in goal-oriented implementation
- Constant analysis of all intermediate results with comparison to the original planning with the help of working memory
- Plan and self-corrections that may be necessary

In fulfilling these frontal lobe tasks, its battery is emptied, which I will refer to in the following as the "frontal lobe battery." As its "charge level" decreases, it becomes increasingly difficult for us to think in a concentrated manner and to muster the motivation to do so. When completely mentally exhausted, it is almost impossible to find the best solution to a problem.

Chocolate and willpower. In 1996, Roy Baumeister devised a psychological experiment that revolutionized our thinking about human willpower.[1] Together with three former colleagues from Case Western Reserve University, he investigated how the unfulfilled desire for chocolate affected the mental performance of his study participants. For the experiment, all participants stayed in a room with freshly baked chocolate chip cookies.

Of course, the smell created a great craving. But while one half was actually allowed to indulge in the cookies, the others only got radishes, for which they could not muster much enthusiasm. Afterward, both groups had to solve difficult puzzles, and something surprising happened: The "radish group" made far fewer attempts at solving the tasks and also spent less than half as long on them as the "chocolate chip cookie group." Having to resist the sweet temptation had obviously cost willpower, which was now lacking in another area. For psychology, this very simple study was a breakthrough and became the basis for thousands of further investigations into the behavior of diverse groups and individuals. All have one conclusion in common: Willpower, or mental energy, is required for numerous tasks and decision-making processes and—this is crucial—it is limited.

Frontal lobe battery depletion affects not only motivation and willpower but also self-control. That's why it also becomes noticeably more difficult toward evening to stick to good intentions. For example, after dinner I find it very difficult to ignore chocolate—even when I'm actually completely full—when I know it is in the house. Nothing stops me from getting it out of the cupboard and then usually eating the whole bar right away—neither my guilty conscience nor the knowledge that I will be annoyed about it the next morning when I grab my belly fat. Apart from the fact that chocolate tastes good and evokes pleasant childhood memories in me, as it does in many people, there are at least three reasons for this behavior with decreasing willpower and self-control or an empty frontal lobe battery, which can be easily explained using this personal example: First, it takes mental energy not to think about chocolate, and this energy is simply no longer available in sufficient quantities in a state of ego depletion. Second, when the battery is empty, one tends to follow spontaneous impulses, such as the idea of being allowed to reward oneself with chocolate for the day's efforts, which is then implemented as quickly as possible. Third, when mentally exhausted, you hardly give any further thought to the longer-term consequences of your actions (too much chocolate is unhealthy in the long run), even if your conscience may bother you at that moment—that would also involve mental effort.

As my beginning anecdote illustrated, I—like you too, probably—discovered early in life that, when your ego is depleted, it's not so easy to be as friendly in conversations as you've learned to be, as you'd genuinely like to be, and as is expected of you. This is easy to understand with regard to Baumeister's chocolate experiment, because after all, our brain's executive center's functions which require mental energy are not only creativity and concentration but also social competence—and this includes learned behaviors or the learned suppression of socially unacceptable reactions. If this ominous energy is used up, then, as a good acquaintance and generally very friendly person once put it to me in a nutshell, what easily happens is: "At some point you just can't be nice anymore."

The common characteristic of all the frontal lobe's activities when thinking, making, and implementing decisions is that they require this special kind of mental energy—regardless of whether you actually do something or seemingly do nothing, even whether you are aware of it or not. This energy is also needed for our willpower, for the ability to keep going even in the face of difficulties and temptations. Persevering and staying focused requires a continuous decision-making process that can deplete our ego. When we no longer have mental energy available to us throughout the day, or in insufficient quantities, when the frontal lobe battery is increasingly "discharged," then creativity, concentration, and self-control also diminish, and we usually make fewer good decisions, or at least not as well-thought-out ones. In a sense, we are like the accident victim Phineas Gage, though we don't suffer from a permanent frontal lobe syndrome—rather an acute frontal lobe impairment.

Fast and Slow Thinking

According to psychologist Daniel Kahneman, how our brain makes decisions and how this process is influenced by ego depletion can be explained by two seemingly competing but ultimately collaborative thinking or decision-making systems: the fast System I and the slow System II.[2]

System I is, in a sense, the autopilot of our brain. According to the "predictive mind theory," System I guides us largely unconsciously through life by constantly predicting what the future might look like and acting accordingly. Since the fast System I, unlike System II, naturally requires very little mental energy, it can be constantly active, and it is. This also makes perfect sense, because in most everyday situations that repeat themselves day after day, it is advantageous to be able to react quickly without having to think first. This applies to routine actions and especially in acutely life-threatening situations, because System I also allows us to make decisions quickly, to speak "from the gut," and to act just as quickly. This is vital for survival, and that's why nature has arranged it so that System I still works even when the mental battery is empty.

The slower, meticulous System II thinking, in contrast to the constantly active System I, is on "standby" and only becomes active when we consciously switch it on. For example, when new life decisions are pending, there are often several alternatives and you feel that you should consider all relevant aspects as much as possible. Then we or our consciousness take control, and the autopilot (System I) is switched off for a short time to conserve our mental battery and to continue to have enough energy reserves for possibly further thought processes using System II. To use System II continuously would very likely not be possible, if only for reasons of capacity. We are by nature in energy-saving mode.

Switching on System II is not so easy, because System I tends to make hasty decisions or hasty answers to questions that seem simple to us at first glance, even if they are not. I would like to illustrate this with the help of the following two examples, which can also be found in Kahneman's book *Thinking, Fast and Slow*.

System I spontaneously responds to the question, "When would a pond be *half* covered with water lilies, if the leaves of the water lilies double every day and the pond would be completely covered after 12 days," with the answer, "after 6 days" (that would be *half* the time). The error in thinking becomes clear as soon as you switch on System II and find the correct answer, "after 11 days" (the opposite of doubling is halving, so counting backward is also one day).

Similarly, the spontaneous answer from System I to the question of what a ballpoint pen costs if it and a watch together are worth €110 and the watch is €100 more expensive than the ballpoint pen is usually "€10" (110 minus 100 is 10). Here, too, System II would realize with a little thought that this was a "not-thought-out error" of System I and that "€5" is the correct solution (5 plus 100 is 105, and 5 plus 105 is then the €110 actually paid).

Interestingly, a correct or incorrect answer says nothing about the respondent's IQ; even people with very high IQs often tend to give the wrong answer. The reason for this is that *all* people usually give priority to System I: on the one hand, because it is the first to "hear" the question, and on the other hand, because it requires less mental energy to answer with System I while System II rests. Only when System I "senses" based on past experience, perhaps because it was once embarrassing to have reacted hastily (i.e., a corresponding gut feeling sets in) and therefore realizes that the answer cannot be so obvious (why else would someone even ask such a supposedly simple question if there wasn't a trick to it?), does our brain switch on System II. However, as ego depletion increases, the likelihood that we will not do this increases, even if our gut feeling advises us to do so.

When we lack mental energy, we are more open to risk and then also quick to give a possibly wrong answer or make a wrong decision. This behavior is reminiscent of the accident victim Phineas Gage, which means that damage to the frontal lobe can have the same effects as frontal lobe battery depletion. If the frontal lobe battery's charge—the energy necessary energy for us to use System II—drops to zero, System I always gains the upper hand in our decisions, and these are usually conservative because remaining in the usual and therefore familiar state requires no reflection and also suggests security. Unfortunately, this also applies when we feel or know that fundamental changes in our behavior or lifestyle would be better in the long run.

This is not only relevant for oneself, since our decisions almost always have consequences for others as well. For example, if I were a prisoner who has served most of my sentence and am rightly hoping for early release due to my exemplary behavior, then an essential precaution would be for my lawyer to ensure that I am the first case the judge hears on the day of the hearing. After all, according to the results of an Israeli study published in a prestigious journal, my chances of being released early into freedom decrease considerably, from a relatively high 65 percent if I am the first to practically zero if I am the judge's last judgment of the day, even if my case is comparable to the first of the day.[3] There are many attempts to explain this phenomenon. However, it is very likely that it is easier for the judge, with increasing fatigue and under the pressure of time (because he wants to go home or have lunch), to be conservative, to leave everything as it is.[4] After all, the decision to release a prisoner early—even a prisoner with very good conduct—requires the consideration and weighing of many more aspects than leaving the current state, which one knows and which harms no one (except the prisoner). Being conservative saves mental energy and time; however, in the example mentioned, only for the judge, not for the prisoner.

Important or unimportant? Everyday decisions that don't feel too important, such as what to put on the shopping list, are made quickly with System I. However, on closer inspection using System II, they could become complex. For example, guided by System I, you would most likely opt for the cheapest piece of meat when shopping, especially if you have internalized the advertising slogan *"Geiz ist geil"* ("Stinginess is cool"), for example. You even feel good about it then. If System II were engaged, it might call to mind the detrimental health aspects, as it's widely known that cheap meat from factory farming (unhealthy fats, antibiotic contamination, etc.) is unhealthy. For emotional reasons, System II might also factor in animals' suffering into purchasing decisions and consider reducing or even eliminating meat consumption. If System II considers the global consequences of meat consumption—e.g. antibiotic resistance due to widespread antibiotics in animal agriculture, environmental damage caused by clearing rainforests to produce animal feed, and groundwater pollution—then a simple choice between cheap meat or a healthy alternative when writing shopping lists becomes a significant decision. It has personal and global consequences, and one should therefore perhaps not leave it solely to System I.

We will return to the problem of conservative thinking and acting in the case of an acutely but also chronically empty frontal lobe battery in more detail, because it has not only personal but also social, health, economic, and even environmental and global political consequences.

Thinking and Acting in Prejudices

System I acts—precisely because it does not expend much mental effort—stereotypically out of necessity. This does not pose a problem in acute emergency situations. On the contrary: If, for example, a car is speeding toward me, it would make little sense to think for a long time about who the driver might be or what he might intend with his driving style, whether he just wants to scare me or whether an accident is actually imminent. Here it is important to avoid the car immediately. "Cars driving toward me are dangerous" is a lifesaving stereotype.

The American writer Walter Lippmann (1889–1974) defined "stereotypical thinking" as thinking in which, for economic reasons, the necessary energy expenditure of a comprehensive analysis of detailed experiences is not undertaken. In other words: Those who think in stereotypes think quickly, but strictly speaking, they do not really think, but rather, according to Lippmann, run the risk of applying "solidified, schematic, objectively largely incorrect cognitive formulas" in everyday life.[5] Furthermore, a stereotype or prejudice acts like a filter that influences one's own perception and thus usually reinforces the existing bias. Thus, one pays more attention to information that fits into a prefabricated scheme than to others. Oxford experimental psychologist Robin Murphy and his team came to the conclusion that our brain prefers to collect negative statements about a group we dislike, while it tends to ignore positive ones.[6] System I is also at work here. The danger of relying only on System I in one's daily decisions increases when the mental battery's charge is either insufficient or no longer sufficient to use System II. For example, prejudices, even if one knows that they are unjustified, can be difficult to suppress in a state of ego depletion.[7]

In these cases, stereotyping becomes a disadvantage. Prejudices such as "dark-skinned foreigners or people of other faiths are a threat" harm a society that relies on global cooperation and ultimately even threaten the existence of all humanity, because unless all people respect each other as equals, global problems that not only affect but even destroy our future prospects cannot be solved.

Good Advice Is Not the Solution

In fact, it's mostly worthless. Since ego depletion is a phenomenon that affects everyone, there is no shortage of general and well-intentioned advice on how to deal with it.[8] It is recommended, for example, to take certain precautions daily as a self-protection mechanism as long as one still has sufficient mental energy. In the case described at the beginning, I asked my assistant not to put any more business calls through in the evening (the line remained open for my family). In my case, it would certainly also be wise not to buy chocolate, so in the evening, when my ego is depleted, I wouldn't be even tempted to eat it.

However, this does not solve the fundamental problem of mental exhaustion. In the first example described, it would have been much better to take an important business call. Avoiding buying chocolate could even make the problem worse (as I also know from my own experience). Because with a depleted ego, I'm apt to rummage through the pantry for other edible alternatives, which might be even unhealthier. Good advice, therefore, doesn't help us much.

It becomes even more problematic when one is at the mercy of another's depleted ego, as illustrated above in the judge's dilemma. This is almost unavoidable in a complex world where everyone is dependent on everyone else in some way. Therefore, it would be better to find a fundamental solution instead of giving advice. This could lie in increasing the mental battery's capacity so that one can do the upcoming day's necessary work without completely exhausting oneself mentally. Thus, everyone—or at least most people—would always have enough mental energy to switch on System II at any time if necessary and act reasonably. A nice thought. But is that actually feasible?

To find out whether an increase in the capacity of our mental energy reserves is even possible and whether this could be achieved naturally, one would first have to find out what exactly *is* depleted during mental exertion, because it is still completely unclear what kind of energy our frontal lobe needs to be able to perform its executive functions by means of System II. The primary question, scientifically posed, is therefore: What is the frontal lobe battery's "neural correlate," or what is the physical nature of the mental energy stored in it? The answer to this question is of such fundamental importance that we will now embark on a search for an answer.

CHAPTER 3

THE SEARCH FOR THE FRONTAL LOBE BATTERY

The measure of intelligence is the ability to change.
—Albert Einstein (1879–1955)

The Sweet Illusion

As I write these lines, I am still in the midst of my nightly fast. Even during my studies, I noticed that I learn and work best when I start right after getting up in the morning without having breakfast first. Even back then, a coffee was enough to get me going, and nothing has changed about that to this day. In my school days, the principle of "breakfast like a king" was downright celebrated. However, I noticed that after eating a jam sandwich or crispy muesli, I would get a queasy feeling within about a quarter of an hour, which manifested itself as ravenous hunger. This seemed paradoxical to me and at the time still inexplicable, because after all, I had just eaten something. Today I know why that was. When sitting, the muscles hardly have to work and therefore have almost no need for energy and thus also not for the carbohydrates eaten at breakfast. Therefore, the blood sugar rises, whereupon the sugar stress hormone insulin lowers it again, causing a slide into hypoglycemia. Even after a sweet snack during school break, my attention was no longer on the teacher, but solely on the clock above the classroom door and the question of whether I would make it to the next break. Sometimes I would secretly open a small square dextrose tablet under the school desk, wrapped in a transparent plastic cover, to solve a problem that shouldn't have existed in the first place. In this way, I made it through the whole day, from meal to snack to meal.

For a long time, neurobiologists suspected that the ego's depletion as a result of mental exertion (System II), the gradual loss of "brain energy," was in fact a problem of energy supply, a deficit of blood or glucose. From this perspective, the previously mentioned study, according to which prisoners are significantly more likely to be released from prison if the judge has had a break beforehand, was also commented on with the somewhat satirical words: "Justice is what the judge had for breakfast."[1] However, this prevailing doctrine at the time contradicts the realization that even with prolonged fasting, very effective mechanisms are set in motion to maintain the performance of the human brain. Thus, the brain is usually by far the last organ to suffer from a lack of energy when fasting: If we do not eat any food for about ten to twelve hours, the blood sugar usually remains within a healthy normal range. In addition, saturated fatty acids can be released from our fat deposits for additional energy supply because no insulin is released in this metabolic situation, which would prevent this. These fatty acids reach the liver via the bloodstream and are converted there into ketone bodies. Ketone bodies—unlike saturated fatty acids—can easily reach the brain and are even a more efficient source of energy there than glucose. Thus, less oxygen is needed when "burning" ketone bodies and yet more energy is released. The idea that ego depletion is a lack of metabolic energy is also contradicted by the finding that more mental work simply increases blood flow to the brain regions that are working harder.[2] Regardless of whether our brain uses the intensively thinking System II or the superficially acting System I in its decisions, there is no reduction in the local blood sugar concentration during mental work.[3] But evidence from another area of brain research also rules out glucose as the limiting factor for brain energy and as the cause of ego depletion. Thus, it could be shown that mental tasks that require self-control and the use of System II are not impaired during simultaneous physical work, even though muscle work consumes significantly more glucose than thinking. Even when blood sugar levels drop slightly due to physical exertion, the exact opposite has been found: Physical activity does not diminish mental performance, it increases it.[4] Thus, in preschool children, even 20 minutes of physical exertion in the aerobic range, in which speaking is still possible, lead to a significant improvement in performance in tasks that require cognitive attention control compared to sitting still.[5] The authors of the study summarized their findings as follows: Overall, these results suggest that a single session of moderate-intensity aerobic exercise (such as walking) improves cognitive control of attention in preschool children and contributes to the use of moderate exercise to increase attention and academic performance. The data suggest that even a single exercise session affects certain processes that support cognitive health and are very likely necessary for effective functioning throughout the lifespan. From an evolutionary biology perspective, this makes perfect sense, because even as the human brain was developing, the gathering of experiences was usually also associated with physical movement such as gathering food. This raises the question of why children have to sit in school when they learn—after all, the point should be for them to stay as focused as possible and open to new knowledge, i.e., with an active System II.

Fasting or physical activity does not restrict thinking, but rather promotes it when the blood sugar is kept in a lower range or even lowered somewhat. The mental energy that System II needs and whose still unknown storage is ultimately depleted must obviously be something other than metabolic energy. Therefore, a supply of glucose does not help when it comes to regenerating the depleted ego.[6] On the contrary: The intake of glucose makes the blood sugar skyrocket, whereupon the body releases large amounts of insulin, which ensures that the glucose is absorbed by fat cells and converted into body fat. As a result, the blood sugar level drops as quickly as it rose before, and usually even more. The production of energy-rich ketone bodies also comes to a standstill. The consequences of the drop in blood sugar are then a drop in performance and cravings, which, if repeated frequently, can lead to weight gain and metabolic disorders. Furthermore, elevated blood sugar levels lead to inflammatory reactions in the brain and, as a result, to poorer memory performance, which is precisely the opposite of what one would actually expect from the sweet energy boost.[7] Nevertheless, almost pure glucose is still sold as mental energy in small, plastic-wrapped dosages, even though one actually knows better. So that parents know right away what their children supposedly need for learning, such glucose tablets bear names like "school supplies." In a corresponding product advertisement it then also says[8]:

> Sparkling, fruity, delicious. Dextro-Energy school supplies in the practical box provides energy and taste at the same time. The fast carbohydrate dispenser for in between and on the go convenience with the aroma of strawberries, raspberries, and blueberries. Dextrose [an old name for glucose] goes directly into the blood [which is true] and promotes concentration and mental performance [which, according to what has been explained before, can only be true if the blood sugar level had previously reached an emergency low, for example in the case of an insulin overdose in a diabetic].

But the high amount of concentrated sugar (83 percent dextrose and maltodextrin) is not the only problem; the other ingredients such as acidifiers, acidity regulators, separating agents, and artificial flavors are food components that are unhealthy for a child (and also for an adult).

Of course, our brain is not efficient without a sufficient supply of metabolic energy in the form of ketone bodies and glucose, which is why an acute circulatory disorder of the brain, such as a stroke, or a chronic one, such as arteriosclerosis, usually leads to a disorder of thinking ability. But even intensive System II thinking neither diminishes the blood supply to the brain (the opposite is the case, as we have seen), nor creates a lack of metabolic energy.[9] In the case of ego depletion and thus the decrease in willpower, it must therefore be a lack of a completely different form of energy. But which one? To get to the bottom of this, we should take a closer look at the elementary properties of this energy or what we know about its storage.

Six Basic Properties of the Frontal Lobe Battery

So far, science has not yet given us an answer to what kind of frontal lobe energy is being depleted nor where this as yet unknown energy is stored in our brain, i.e. where the frontal lobe battery is located. But there are some interesting and promising clues. We know at least that the frontal lobe battery has the following important properties:

1. It is needed for mental work using System II.
2. If the frontal lobe battery is not located in the frontal lobe itself, but in another area of the brain, there must be a fast and direct connection for energy transfer.
3. Its storage is limited, which explains the ego's depletion through its use.
4. In stereotypical activities using System I, it is not used or at least used only to a small extent.
5. It is "recharged" during sleep.
6. Insufficient charging or neurological damage to the frontal lobe battery causes the same symptoms as damage to the frontal lobe.

These properties are also essential identification criteria that will help us in our search for the frontal lobe battery. However, for a reliable identification, all of these criteria must be met.

To get a little closer to identifying the frontal lobe battery, we begin with the question of what our brain has to do when we think intensively with System II and possibly even decide to break new ground instead of acting habitually with the conservative System I. What is special about System II thinking?

Intellegere—Deciding Based on Memories

Usually, there are at least two possibilities at the forks in our life's path. If we want to make a responsible decision in one direction or the other at such a point, we should mentally play through each option in terms of its immediate but also its long-term consequences. In doing so, we could, depending on the given situation, ask ourselves the following questions: What effects will my possible action or inaction have on my life, but also on the lives of others who might also be affected by my decision? What emotions might be, or are very likely to be, evoked in those involved by the respective alternative? This is also important. Because the calculation of emotions, feelings and affects, even in a supposedly purely rational decision-making process, is not a dispensable luxury—on the contrary. For example, if we do something that harms people in our personal environment who are important to us, social isolation and loneliness could be a consequence, which, at least in the long run, would also significantly damage our own well-being and health. With some decisions, we would also have to think about what the consequences of our actions could be on the environment, for example. It would be important to factor this in, especially with many everyday tasks, precisely because they have often become routine. Because what we do every day is very likely to have a greater impact on our environment than one or the other one-off action (exceptions prove the rule here). For example, a one-time flight will cause significantly less environmental damage than the daily consumption of meat from factory farming. Since the latter is also directly detrimental to one's own health, one might first consider alternatives. So our actions are often also about living up to our responsibility for ourselves, but also for our immediate environment or even for society as a whole.

Making the most balanced decisions possible is not easy and requires a certain degree of intelligence. This is already suggested by the original meaning of the term, which comes from the Latin verb *intellegere*, a combination of the words *inter* for "between" and *legere* for "choose," and thus means "to choose between at least two alternatives." Intelligence is often equated with the result of an intelligence test. However, this only determines the IQ, the intelligence quotient. Its level correlates with the performance of the working memory, which is determined by the number of information units that one can mentally "juggle" at the same time during problem-solving. Five to six things represent a good average here. The problem with this: When juggling balls, they only stay in the air as long as you consciously keep them moving. If the juggler is distracted for just a moment, they threaten to fall down. Even when juggling our thoughts in working memory, they are only present in our mind as long as we consciously keep the brain waves that represent them active by concentrating on them. If our attention is distracted for even a small moment, the respective train of thought is interrupted. When doing mental arithmetic, for example, any intermediate results only remain in the working memory for a few moments, ideally just as long as we need them for the calculation. As soon as the result is available and our mind wanders to the next topic, they are lost. Also, an unknown telephone number that we have memorized for the purpose of typing will usually be forgotten immediately as soon as the conversation starts. The working memory thus contains only fleeting information that, as with a computer, is lost as soon as we switch it off without having saved the data beforehand. Due to the volatility of the information, working memory is also called short-term memory.

The level of IQ or the strength of working memory, therefore, does not help us in the search for the frontal lobe battery. After all, decisions using System II, weighing the advantages and disadvantages of all options, depend on our being able to consider very many aspects and also remember our thoughts in the longer term. Since the working memory is obviously not capable of this, it cannot be a candidate for the sought-after frontal lobe battery.

A large working memory and, accordingly, a generally high IQ are also no guarantee we will actually use System II to solve a problem. As the two examples in the previous chapter show (growth of water lilies and cost of the ballpoint pen), even a high IQ does not protect against making a possibly wrong decision hastily with System I. The level of IQ does not guarantee that one uses the capacity of one's working memory for the benefit of society. The decision, for example, whether to contribute to the solution of an impending environmental catastrophe through one's actions or to simply take the most convenient path to enrich oneself, without considering the long-term consequences for the environment or posterity, is completely independent of IQ. In this respect, it can be compared to the power of a car. Here, too, the number of horsepower is independent of their use. A high horsepower number can be used for a lifesaving drive to the hospital, but in the other extreme also for a deadly rampage into a crowd of people. Due to such considerations, the term "intelligence" is rightly no longer equated with IQ, but rather corresponds much more to a combination of rational (RQ), emotional (EQ), and social (SQ) intelligences. It immediately becomes obvious that the "fluid" (because it is constantly in flux) short-term memory, and thus also the IQ, cannot even come close to being sufficient to take into account rational but also social and emotional parameters. Our thoughts and insights that arise during reflection using System II must be stored simultaneously and efficiently in order to be able to retrieve them just as efficiently. Likewise, for important decisions, all relevant previous life experiences should be quickly recallable. In short, System II uses fluid working memory to form thoughts and keep them active in consciousness, but it must also be able to simultaneously access a much more permanent memory.

In contrast to fluid (unstable) short-term memory, System II requires "crystalline" (stable) long-term memory. This must be able to permanently store different future scenarios, even while we are playing them through in short-term or working memory. This must happen instantaneously, because every thought that develops in working memory could be important and must therefore not be lost. Only with the help of such a memory system can we compare the action options played through in the context of a decision-making process. Every person knows from their own experience that such a crystalline memory (on which the so-called crystalline intelligence is also based, which can also be described as experience-based knowledge or wisdom of old age) actually exists: Often we can still remember years later what went through our heads during a conversation or when making a decision, and especially what emotions accompanied these thoughts. A memory that quickly and permanently captures our thoughts and experiences with others is also a prerequisite for a functioning social life. After all, one can only interact meaningfully and reliably with one's environment if one remembers past interactions. Intelligence is therefore the ability to create new memories from old ones. The long-term memory required for this must then store our intention and the associated plan so that we can also put it into practice. Such a fast-learning and stable long-term memory is needed for mental work using System II.

The Hippocampal Memory Store

Without a long-term memory that stores our personal experiences and especially our thoughts, System II cannot function. However, for a long time, it seemed to science to be an almost unsolvable mystery where in the depths of the human brain our thoughts, conversations, and in general everything that happens in our lives, all our individual experiences (which make us individuals in the first place), are stored. A pioneer in the search for this place of our personal memory was the American psychologist Karl Lashley (1890–1958). His working hypothesis was as simple as it was crude: You have to find the memory by destroying it. Tirelessly and very systematically, Lashley removed smaller and larger parts of the brains of thousands of experimental animals through surgical procedures. But at the end of his research career, he was baffled to find himself empty-handed, because not a single part of the brain seemed to harbor personal memories. So Lashley drew the depressing conclusion: "This series of experiments provided (indeed) a large amount of information [about all sorts of functions of the brain], but not about where our memories are to be found."[10]

What Lashley had not succeeded in doing in decades of meticulously planned scientific work, the neurosurgeon William B. Scoville (1906–1984) achieved on September 1, 1953—albeit completely unintentionally.[11] Perhaps Scoville was motivated in his approach by the "lobotomy Nobel Prize," which had been awarded only a few years earlier (this seemed to allow destructive interventions in the human brain). In any case, Scoville's urge to experiment was probably greater than his concern about possible consequences. Since Scoville had no idea what functions of the brain he would destroy with his intervention, the result of the operation was as surprising for him as it was tragic for his patient, the then twenty-seven-year-old Henry Gustav Molaison (1926–2008).

Molaison had fallen off his bicycle as a child. Since then, a small scar in his brain had caused spontaneous electrical short circuits. Over the years, these developed into life-threatening epileptic seizures. To solve the problem once and for all, Scoville decided to eliminate the source of the electrical discharges. However, perhaps due to an aesthetic penchant for symmetry, he did not just remove the scarred tissue on one side of Molaison's brain. Instead, he removed, axisymmetrically and quite generously, a thumb-sized structure located deep in each temporal lobe. This resembles a seahorse in its anatomical form and is therefore called the hippocampus.

Figure 3: The hippocampus, a brain structure that looks similar to a seahorse.

The case of Henry Molaison, deprived of his two hippocampi, revolutionized brain research and confirmed Karl Lashley's working hypothesis: The personal memory had been found by destroying it. Henry Molaison lost his individual memories because of the operation. All previous episodes of his life were erased, which is why the hippocampus, or the two hippocampi, are referred to as episodic memory storage. Crucial to Scoville's discovery of episodic memory was the fact that he destroyed both hippocampi, which complement each other in their function, at the same time. This was the only way Molaison could actually lose the ability to remember past experiences or memorize new ones.

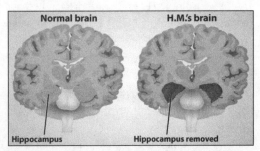

Figure 4: Henry Molaison lived another fifty-five years after the removal of his two hippocampi. To keep his identity secret during his lifetime, his long and extensively documented medical history was published under the initials H.M.

All other forms of memory, such as language, learned factual knowledge, and motor skills, were retained by Molaison. His working memory, also located in the frontal lobe, remained intact, which was why his above-average IQ did not change after the operation.[12] He was thus still able to retain thoughts for a short time through concentration, even if he forgot them forever afterward. This also allowed him to have conversations. However, as soon as he was distracted, he remembered neither their content nor the person he had been talking to.

By storing our important thoughts, conversations, and events in the long term, the episodic memory spins a common thread through our existence and gives the feeling of continuity. Ultimately, it gives us our identity, which is also defined by our personal relationships with our environment. This too was taken from Molaison. According to Brenda Milner, a British Canadian neuropsychologist who looked after him for many years and revolutionized cognitive neuroscience through her studies on him, he lost something crucial through the loss of his hippocampi, namely his humanity. Although he was always good-natured and friendly, one could not, as she felt, "develop a friendship or any human affection for such a person."[13] This may sound harsh, but it is understandable, because without functioning hippocampi, it is not possible to remember personal experiences and thus also not possible to build social relationships.

Henry Molaison was, after the loss of both his hippocampi, a man without plans, goals, or hopes, for what memories were there to base these on? For him, thinking and planning using System II was no longer possible. After all, episodic memory is the place where the consequences of possible actions, which we analyze with the help of the working memory in the frontal lobe, are stored quickly and effectively. In this way, we can later weigh them up against each other in the frontal lobe in order to make decisions and implement them. Looking at these properties, episodic memory meets our first criterion for the frontal lobe battery: "It is needed for mental work using System II."

Cerebral Information Highway

In order to be able to make expedient decisions, the executive center of our brain needs current information from the environment via the sensory organs. But it must also have access to all previous experiences that could be relevant to the given situation. Furthermore, all current thoughts or new ideas that are important for decision-making but cannot be simultaneously active in the fluid working memory due to capacity reasons (because only a few mental contents can be "juggled" there at the same time) must be retrievable. And ultimately, all this different information must reach the frontal lobe quickly. To ensure all this, the prefrontal cortex is directly connected to all brain regions that are needed in these complex thought processes (System II). This happens via a network formed from extraordinary nerve cells, the Von Economo neurons (VENs).[14] The VENs are named after their discoverer, the Greek psychiatrist and neurologist Constantin Alexander Freiherr Economo von San Serff (1876–1931). They are characterized by their extraordinary size and concomitant high speed in the transmission of signals.[15] In contrast to the other classes of nerve cells, VENs in humans only develop after birth, but then their number increases steadily until the age of four. VENs play a central role in the integration of body feelings, in the emotional evaluation of social situations, and in goal-directed cooperative behavior.[16] In particular, socially relevant experiences are "delivered" from the hippocampus to the frontal lobe via the VENs, which form a kind of information highway, so that it can perform its executive functions quickly and effectively in socially tense situations.[17] Studies have shown that the hippocampus itself actively contributes to problem-solving in conflict situations. It is particularly active in people who succeed quickly and successfully.[18] VENs are therefore indispensable for the intuitive assessment of emotionally charged situations and enable our executive center, the frontal lobe, to adapt our behavior to rapidly changing social situations. VENs very likely already gave our prehistoric ancestors a survival advantage; they could use them to judge in a matter of seconds who they could trust based on their behavior and who they could not. Higher primates, elephants, and marine mammals, all of which are notable for their outstanding social skills, also possess this special "wiring." VENs are certainly one of, if not *the* decisive prerequisite for these animals to have the same capacity for empathy as humans and also the gift of social and planning intelligence.

Some studies have revealed a reduced number of VENs in people with certain neuro-psychiatric disorders in which social behavior is always significantly impaired. An example would be frontotemporal dementia, in which these nerve cells gradually and selectively die for unexplained reasons.[19] As a result, System II fails or can at least no longer be used effectively, especially since the crystalline memory of the hippocampus can no longer send socially relevant data to the frontal lobe. Typically, those affected become increasingly superficial, inattentive, act rashly, and neglect their duties even at the beginning of the disease. At work, they usually attract attention due to repeated mistakes. Also striking is an increasing tactlessness in dealing with their fellow human beings, an increased irritability, and in some cases also an increased aggressiveness, even if the affected persons were very calm and friendly people all their lives. As a result of the massive disruption of System II, they talk "off the cuff," without thinking or reflecting on the effects and consequences of what they say beforehand. They do not take social norms into account. Instead, they uninhibitedly express their stereotypical thoughts, often in an extremely rigid tone. They are thus extreme examples of purely System I–controlled persons.

With the progression of frontotemporal dementia, the phase of withdrawal follows. Then patients lose interest in their fellow human beings, becoming indifferent and listless and finally completely apathetic. With the collapse of the connection between the frontal lobe and the hippocampi (and ultimately also their destruction), patients often develop an excessive hunger, especially for sweets—not because the frontal lobe lacks energy, but because the loss of hippocampal influence on frontal lobe function means that the long-term health consequences of their actions are no longer given any importance. This is also contributed to by the fact that any willpower is gradually lost.

Although frontotemporal dementia represents an extreme example of stereotypical behavior, it nevertheless provides further evidence of the importance of the hippocampus for the healthy executive function and performance of the frontal lobe. To significantly restrict the frontal lobe in its function, it is sufficient, as this clinical picture shows, to selectively destroy the VENs, i.e., to sever the conduction pathways between the two brain centers. This also happened, albeit surgically, in the previously mentioned Nobel Prize–worthy lobotomy.

This Nobel Prize was awarded in 1949 to the Portuguese neurologist António E. Moniz (1874–1955) "for his discovery of the therapeutic value of prefrontal leucotomy [the severing of fiber connections of the central nervous system; a synonym for lobotomy] in certain psychoses."[20] The American physician and psychiatrist Walter J. Freeman (1895–1972), also known as "the man with the ice pick," became a convinced advocate of lobotomy.

According to research by the science magazine *GEO*, medical records in the US between 1940 and 1944 record about 684 lobotomies.[21] Walter Freeman himself had operated in over 200 cases by 1943, with a "success rate" that he put at just over 60 percent. Freeman was not only enthusiastic, he also had a mission, because at the end of the Second World War the country's state psychiatric hospitals were filled with hundreds of thousands of patients and completely overwhelmed. His efficient ice pick technique of lobotomy was supposed to solve the problem and not be the last therapeutic step for these patients, but the first. That's why Freeman "industrialized" Moniz's original method into a mere seven-minute procedure under local anesthesia (the brain itself feels no pain). Thousands were lobotomized annually thereafter, and tens of thousands after lobotomy was honored with the Nobel Prize. Later, Freeman wrote self-critically about this highly controversial type of psychosurgery, saying that it achieved its supposed successes by "shattering imagination, dulling emotions, destroying abstract thinking, and creating a robot-like, controllable individual."[22] As early as 1950, the then Soviet Union banned lobotomy because it "contradicts the principles of humanity." Soon after, countries such as Germany and Japan followed suit.[23]

Both the early symptoms of frontotemporal dementia and the devastating consequences of lobotomy show that the collapse of data flow to the frontal lobe has the same effects as a direct dysfunction of the frontal lobe itself. In particular, there is a complete loss of function of System II, as can be concluded from Freeman's description of the consequences of his lobotomy. Since the hippocampus is needed for System II use, the second criterion would thus also be fulfilled: "If the frontal lobe battery is not located in the frontal lobe itself, but in another area of the brain, then there must be a fast and direct connection for the energy transfer." Frontotemporal dementia in particular suggests that VENs could play a role, because both hippocampi have a direct and extremely efficient connection to the frontal lobe for "energy transfer" through them.

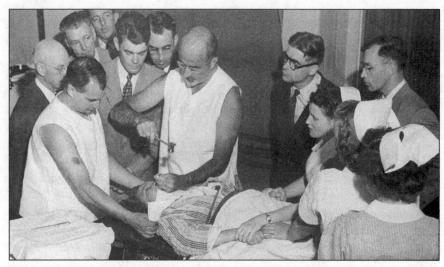

Figure 5: The man with the ice pick. With no regard for hygiene, the "neurosurgeon" Walter Freeman pierced through the eye socket and into the brain of the awake patient with an "ice pick" that he held in his bare hand to destroy the patient's prefrontal lobe pathways. His spectacular approach attracted onlookers.

Limitations of Hippocampal Memory

The use of System II requires fast and efficient storage of our thoughts. The hippocampus is responsible for this. Its storage capacity is determined by the number of its nerve cells and synapses, whereby the following applies: The larger its volume, the more extensive its memory.[24] It is therefore to be assumed that intensive use of System II pushes the hippocampus's storage space to its limits. Thus, the hippocampus also fulfills the third criterion of the sought-after frontal lobe battery: "Its memory is limited, which explains the ego's depletion through its use." If the hippocampal thought memory is full, System II is no longer usable, which would be equivalent to the ego's depletion. This also fits with the finding that in ego depletion the ability to retrieve specific memories via the episodic-emotional memory, through the hippocampus, is significantly reduced.[25]

The Hippocampus Stores Only What Is Worth Remembering

To ensure that an overload of the hippocampus and thus a System II functional loss of the frontal lobe does not occur prematurely, before the respective day is over, nature has built in an effective mechanism that, so to speak, separates the wheat from the chaff. To do this, the hippocampus uses an ingenious filter: emotions.[26] Only experiences and thoughts that touch us in some way are stored. This makes sense not only for reasons of space, because compared to, for example, trivial, constantly recurring action sequences, emotionally significant memories are usually more likely to be of long-term or possibly even of vital importance. The hippocampus only makes these available to the frontal lobe (and thus also to us) as a memory and basis for decision-making. Conversely, this consideration makes it clear that the hippocampus is neither needed nor used for stereotypical (repetitive) thinking by means of System I. In addition, constantly recurring thoughts, even if they should be emotionally charged, are not stored by the hippocampus as individual events and thus consume basically no storage space. If the hippocampus is indeed the sought-after frontal lobe battery, it would also fulfill the fourth criterion: "In stereotypical activities using System I, the frontal lobe battery is not used or at least only to a small extent."

The Hippocampus Can Regenerate Its Memory Storage Capacity

If the hippocampus is indeed the sought-after frontal lobe battery, then "discharging" would most likely correspond to the gradual loss of its limited capacity to store new thoughts and experiences. If this were the case, however, it would still have to be clarified what exactly is "depleted" by its use and what exactly happens during "recharging." This will be the subject of the next chapter. But the following should be mentioned here: A great many different, successive, and interlocking mechanisms ensure that the hippocampus regenerates its memory store during deep sleep without the stored contents being lost. The hippocampus can only support thinking using System II again after a restful sleep. Thus, it also fulfills the fifth criterion: "The frontal lobe battery is recharged during sleep."

Motivation and Willpower Based on Hippocampal Remembering

Without memory of our past, we can neither plan nor shape our future, as the case of Henry Molaison tragically shows. For him, there was nothing left at all by which he could have oriented his life. "Without instructions," as journalist Philip Hilts, who interviewed Molaison several times for his biography,[27] describes him, "Henry was free of any motivation." As we now know, our motives for action and thus also our plans and goals arise in the frontal lobe, but always on the basis of the thoughts and experiences stored in the hippocampus. Regarding this aspect, there was another attempt a few years ago to explain the unsolved problem of what exactly happens when the ego is depleted. It was assumed that if our brain lacks the energy to use System II in decisions, a lack of motivation could also be the reason from a psychological point of view.[28] This, too, could indicate that the hippocampus serves as the frontal lobe battery; after all, studies have shown that a larger hippocampus leads to a higher motivation to have new experiences.[29] People with a weakened hippocampus are hardly able to try something new and change their habitual life patterns. This is true even if they realize on a rational level that these patterns are objectively harmful to themselves or others. In addition, without a functioning hippocampus, set goals are also easily forgotten.

Due to these and some other important aspects, which we will discuss later, the hippocampus thus also fulfills the sixth criterion on our list of properties of the frontal lobe battery: "Insufficient charging or neurological damage to the frontal lobe battery causes the same symptoms as damage to the frontal lobe itself." These include all symptoms of ego depletion, such as a lack of motivation and willpower. The hippocampus is thus the best candidate for the sought-after frontal lobe battery; no other part of our brain fulfills all six functional criteria. But what is the relationship between the hippocampal thought memory and the memory for the mental energy that System II thinking requires?

Information, a Special Kind of Energy

A battery is, in the technical sense, a rechargeable system that stores electrical energy on a chemical basis. The hippocampus also stores energy, but of a mental nature. By accumulating or collecting information, it enables the frontal lobe to think and act in a complex and planned way by means of System II.

Interestingly, there is a school of thought in modern physics according to which everything that exists consists of information. For this school of thought, the American theoretical physicist John A. Wheeler (1911–2008) coined the expression "It from Bit."[30] According to Wheeler, this expression symbolizes the idea that every object in the physical world has at its core—and indeed at its deepest core—an immaterial source and explanation. What we call and experience as reality (as the *It*) ultimately results from the asking and answering of digital yes-no questions (*Bit*). In short: According to Wheeler (and many other physicists), all physical things could have an information-theoretic origin. In fact, information has already been experimentally transformed into energy, which seems to confirm Wheeler's thesis.[31]

Against the background of this principle, according to which information can also be a kind of energy, the question now arises in a very practical way how mental energy, which the frontal lobe needs, is generated in the hippocampus. This is the question of the actual energy or information carrier.

CHAPTER 4

THE NATURE OF
MENTAL ENERGY

Follow your dreams, they know the way.
—Unknown

Mental Capacity Limits

Right at the beginning of my medical studies, I reached the limits of my mental capacity. Medical knowledge was like the often-quoted bottomless barrel for me. For every extensive book, there was an even more extensive one, and on top of that, the countless new scientific papers! Knowledge was increasing faster than it could be learned. As an autodidact, my main occupation was reading. But the more I read every day, sometimes ten to twelve hours, the less seemed to stick. As the day went on, my brain felt more and more like a table overflowing with goods, with no room for anything else, and from which the new things I was still trying to pile on immediately fell off again, along with the things I had already placed on it before. I felt like I was forgetting more than I was learning. This way of working was ineffective and, above all, exhausting. It soon became clear to me that I had to develop a new learning strategy, otherwise burnout (a term that was not yet common at the time) would only be a matter of time. I started to observe my learning. I found that I could remember well what I had read in the first hour in the morning. What I had read in the second and partly also in the third hour was still largely retrievable the next day. But all further hours spent learning seemed to be a pure waste of time for me. From this realization, a simple plan developed. Since my daily effective learning time was obviously limited, I restricted it to three hours in the morning and possibly another hour after a short nap. From then on, I learned seven days a week, including weekends. The approximately four hours per day, extrapolated throughout the week, were significantly less

29

*than the previous ten hours for five days, but much more stuck, and I also had more time
for sports and other hobbies.*

Of all the regions of the brain, the hippocampus and the olfactory brain are the only
ones that can learn quickly or instantaneously. Once thought, seen, heard, or even
smelled, already a lifelong memory can be created. Both brain regions belong to the
"original" cerebral cortex, the archicortex. *Original* because these brain areas arose with
the evolution of the first vertebrates, very early on, long before the development of
mammals, which in turn are distinguished by the development of the neocortex or the
"new" cerebral cortex.

Even for the first vertebrates, it was vital to have a fast-learning archicortex. In
mammals, and thus also in us, it is not the neocortex, but only the archicortex that
has this unique property of being able to permanently remember something that one
has experienced or even thought only once. This ability makes the hippocampus, in
particular, the memory center for all our personal memories and the guardian of our
very personal life story. It is thus also the part of our brain that gives us our identity.
This also makes it the ideal intermediate storage for thought processes using System
II: It is able to continuously create "backup copies" of all thoughts that, in the working or short-term memory of the frontal lobe, are "juggled." The storage takes place in
the highly flexible connections between its nerve cells, the synapses. *Synapse* is a word
synthesis from the Greek words *syn* for "together" and *haptein* for "grasp" or "touch."
Synapses are thus the places where nerve cells approach each other. They are also the
place where information or memories are stored. This works as follows: Every exchange
of information between two nerve cells leads to a lasting change in the hippocampal
synapse used in the process. In most cases, the synapses become more conductive and
thus "remember" better the information they originally transmitted. In this context,
one speaks of an increase in synapse strength. In neuroscientific terms, the process is
called long-term potentiation because the synapses become more easily activatable for
a longer period of time and are thus, in a certain way, more potent. The synapses of
the hippocampus—compared to the synapses in other brain regions; the only exception is the olfactory brain—can be particularly well modified by long-term potentiation. This predisposes the hippocampus to quickly store our unique thoughts and daily
experiences. In contrast, our motor memory, for example, is based on comparatively
slow-learning synapses. That's why it usually takes many repetitions to learn certain
movement patterns, such as juggling or playing the piano.

We also learn languages slowly, because the change in synapse strength in the language center takes place only in very small steps. If we could learn languages by means
of the hippocampus, hearing or reading once would be enough to learn a new word
or sentence. This is actually possible, but only if we "trick" the hippocampus and learn
vocabulary associatively. Individual facts, such as vocabulary, are uninteresting to it, but

it remembers connections very effectively. The Swedish military, for example, used this to teach elite soldiers without any previous knowledge a difficult foreign language such as Arabic or Russian to the level of an interpreter in just thirteen months. The trick: The soldiers learned so quickly because they linked the learning content with physical activities (more on that later) and emotional experiences, thus making it interesting for the hippocampus. This means that we can also learn facts like foreign words hippocampally and thus extremely quickly, but only if they are experienced emotionally. In other words: It is easier to learn with fun (with sad experiences initially also, but in the long run sadness damages the hippocampus). In a corresponding study, a considerable enlargement of the hippocampus of the recruits was already observed after three months, and in proportion to the respective language acquisition, which ultimately means that the hippocampus grows with its task.[1]

The increase in synapse strength through long-term potentiation or remembering occurs in the hippocampus with the help of glutamate. This is a messenger substance that is released by most active hippocampal synapses for information transmission. Glutamate is therefore also called a neurotransmitter. However, a single synapse altered by glutamate cannot store a thought, an experience, or the content of a conversation on its own. Rather, each memory consists of a combination of many hippocampal synapses that were activated during an event or thought and have increased their strength.

But now we come to the described image of the feeling of the filling and ultimately overflowing table when accumulating new facts or also remembered experiences or thoughts. How does the hippocampus ensure that new memories do not push earlier ones "off the table"? Well, synapses remember remembering, and they do this in the following way: So that, for example, the content of a conversation or thought, let's call it event B, does not "overwrite" a previously stored event A by using the same synapses again to remember a new experience, each long-term potentiation by glutamate is accompanied by the release of β-amyloid. This small protein (also called a peptide) inhibits the further release of the messenger glutamate and thus a further long-term potentiation of the corresponding synapse.[2] Through the released β-amyloid, the synapse "remembers" that it has already stored a new memory. The β-amyloid thus directly protects our memories. At the same time, it indirectly ensures that new events that are to be remembered can only be stored by other, still free synapses. This peptide fulfills in a certain way the same function as turning the page for entries in a diary; it ensures that a fresh page is always written on and that an already used one does not become illegible by a further overwriting. However, this protection of our new memories comes at a price. In the hippocampus, β-amyloid accumulates more and more during the course of the day, while at the same time the number of "free" synapses that could store experiences, conversations, or new thoughts by means of long-term potentiation decreases more and more. This is understandable; a diary does not have an infinite number of pages, and by writing on each new page, the remaining capacity is reduced. This shows that the hippocampal synapses fulfill the third criterion of the frontal lobe battery: The number of free synapses and thus the storage capacity is reduced by intensive mental work. The ego's depletion or mental exhaustion would thus be the exhausting of the hippocampus's storage capacity, a decrease in the free hippocampal synapses capable of storage.

The fast-learning hippocampal synapses thus represent the "mental energy" that our frontal lobe needs in order to be able to think, make plans, store them, and implement them by means of System II. Without free synapses for further thoughts that would have to be stored, the ego is depleted. Then only routine thoughts or routine actions are possible, which do not have to be remembered. The frontal lobe battery's capacity thus corresponds to the sum of all synapses that can be used to store thoughts. Its charge state at any given time is the sum of the synapses not yet used for storage. So when we think using System II, learn intensively or mentally exert ourselves a lot, or simply experience a lot emotionally, many synapses are occupied very quickly. The frontal lobe battery is then, depending on its original storage capacity, possibly empty after a short time because no free synapses are available anymore. Further use of System II as well as any effective remembering would only be possible if previously stored contents were overwritten despite β-amyloid blockade (like the overflowing table on which new things can only be placed if the ones already on it fall off). But only when we sleep does the mental energy miraculously return. The question now arises as to how the frontal lobe battery is recharged through sleep. As is now known, the recharging of the frontal lobe battery takes place in four steps, some of which follow one another and some of which overlap.

Step 1—Upload of Hippocampal Storage Contents

First, during deep sleep, all experiences and thoughts stored by the hippocampus during the course of the day are uploaded to the neocortex. This is possible because the neocortex—in contrast to the hippocampus—has an enormous storage capacity. However, since the synapses of the neocortex learn only very slowly, the hippocampus becomes a tireless storyteller during slow-wave sleep (SWS), a repetitive deep sleep that predominantly takes place in the first half of the night. During SWS, the hippocampus communicates all thoughts and impressions to the neocortex until they are firmly stored there. Apart from the repetitive nature of the data transfer (the "upload," see Figure 6), the entire procedure is similar to that of a digital camera, where the images are uploaded from the full memory card to a hard drive in order to secure them for long term and to be able to take new pictures again. The only thing that remains in the hippocampus are the "mailing addresses," that is, the information with which the memories can later be retrieved. This "postal" information is the location and time coordinates of the memory. That's why it's easier for us to remember things later if we recall in our memory (the "download," also see Figure 6) when and where we experienced or thought something. The memory for space and time, in fact, remains in the hippocampus for a lifetime, while the content of the experience can only be found in the neocortex.

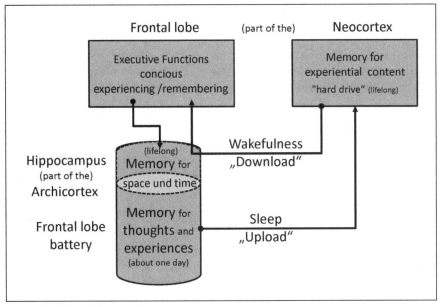

Figure 6: Upload during sleep.

However, a memory is not stored in a single location, like a photo file on a computer's hard drive, but fragmented and distributed across the neocortex. This storage in fragments, in addition to the fact that there are two hippocampi, is one reason why Karl Lashley could not be successful. By removing individual parts of the neocortex (or only a single hippocampus), a memory can never be completely erased.

Emotional memory and emotion reactor. Deep inside the brain, in the anterior part of the temporal lobe between the frontal lobe and the hippocampi, are located, again in pairs, the two amygdalae. They form a part of the brain that we have in common with all mammals. The amygdalae remember the feelings that accompanied experiences.[3] As soon as one finds oneself in a similar situation, the amygdalae trigger the appropriate emotional reaction at lightning speed. If, for example, one was barked at violently or even bitten by a dog as a child, then the barking or any sight of a dog will cause an increased release of stress hormones. The amygdalae are the center for remembered fear.

Step 2—Wisdom through Sleep

In the second half of the night, the freshly uploaded memory contents are related to earlier ones: We dream. This leads to our eyes sometimes moving rapidly as we wander through our dream world, even though we are actually sleeping very deeply. This type of deep sleep is therefore called rapid eye movement, or REM sleep for short. During REM sleep, our brain processes all of our emotionally significant experiences and thoughts of the previous day and weaves them with previous memories into individual insights. This is the crucial reason why we usually become "wiser" in our sleep.[4] Because this gain in knowledge happens during sleep, the time we spend sleeping and especially dreaming has an elementary significance for our mental maturation that goes far beyond recovery.[5] The hippocampus does not record anything during deep sleep, so we only remember dreams if we wake up directly from them. Only when we are awakened do the last contents of the dream become conscious and become memories. That is why one always remembers only the end of a dream, but never the beginning. In addition, most dreams that we remember are nightmares because they are more likely to wake us up than quiet dreams.

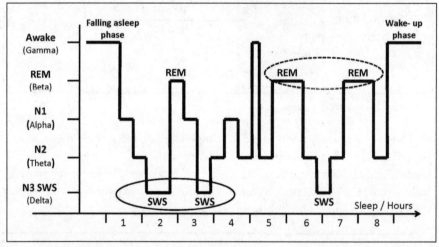

Figure 7: Sleeping in waves.

We enter deep sleep through various stages: First, via the relaxing alpha waves (phase N1) to the theta waves (phase N2), which with frequencies between 4 and 8 hertz dominate the first, still light sleep phase. This is followed by a dreamless deep sleep, which is characterized by delta waves (phase N3). They have the lowest frequency, with only 0.1 to a maximum of 4 Hz. Because of these slow waves, this type of deep sleep is called slow-wave sleep. The learning that takes place during sleep is profound, complex, and at the same time fundamental to the very likely unique ability of humans to see any experience in the context of a larger whole, to draw conclusions, and to change themselves and their environment. Sleep is thus the engine of cultural creation. But it is also crucial for us to be able to pass on our individual experiences to the next generations. For it is only through "uploading" into the neocortex that remembering becomes long-term. Basically, the following applies: The stronger the emotional quality of what was experienced or thought, the more strongly the amygdalae (emotion reactors) were or are active during the original experience or remembering, the more likely the memory of it will be preserved, and the more life-changing it will be.[6]

Gut feeling through dreaming. If previous insights are linked with new experiences or fresh memories in the dream phase, new insights emerge. However, compared to concrete memories, these have neither place nor time coordinates with which one could associate them. Therefore, we cannot know exactly how many and which actual events were interwoven into such superordinate experiences. And so what usually remains is a gut feeling, a reliving of the emotions that underlie these memories. Even if such a gut feeling rarely reminds us of a specific event, because often many events contribute to it or happen in a way that makes it difficult to link them, it nevertheless influences our future decisions. Its contribution should not be underestimated; gut feelings often consist of many individual experiences at the same time.

Step 3—Restoring Memory Potential

After uploading the new memories from the hippocampal archicortex to the neocortex, the β-amyloid previously released at the synapses must be disposed of to recharge the frontal lobe battery. Since it prevents the overwriting of previous memories, the hippocampus would otherwise no longer be receptive. This is again comparable to a digital camera: Here, too, after the old photos have been uploaded to a hard drive, they must be deleted from the memory card. Only then is further photographing possible. For the same reason, during REM sleep (i.e., after the data has been uploaded and thus secured), the β-amyloid is "washed out" from the hippocampus and ultimately from the entire brain. It is then largely broken down in the liver. The lymphatic system is responsible for the β-amyloid transport from the synapses into the bloodstream.[7] This is a drainage system specially developed for the brain, which all vertebrates possess and with the help of which the space around the nerve cells is cleaned of waste products. Without REM sleep, the β-amyloid would remain in the synapses and interfere with or even prevent the storage of new thoughts and experiences the next day. This explains why we have memory and attention problems if we have slept poorly the night before.

Furthermore, the synapse strengths are reduced in the hippocampus.[8] This is also very important for the recharging process because if synapses were to continue to increase their strength with every memory, no long-term potentiation would be possible beyond a certain threshold, and we would no longer be able to remember new thoughts or experiences. In addition, the hippocampal nerve cells would also run the risk of being overexcited and destroyed at too high a stimulus threshold due to an excessive release of the messenger glutamate.[9] In this context, one speaks of the so-called glutamate toxicity.

Both the breakdown of β-amyloid and the reduction of synapse strengths thus ensure that the frontal lobe battery is recharged and ready for action the next day when we are rested.

Step 4—Growth of the Frontal Lobe Battery

In order to "recall" the memories stored on the neocortical "hard drive," the hippocampus is still needed. After all, only it has the respective "postal addresses" under which the fragments of the neocortical memories can be found again. Without a functioning hippocampus, new memories cannot be formed and previously stored ones cannot be retrieved. This was also Henry Molaison's problem. His neocortex certainly housed a rich treasure trove of thoughts and personal experiences that had been stored before his fateful operation, but without a hippocampus he could no longer retrieve them, not even with the help of his prefrontal cortex, even though it itself is part of the neocortex (see also Figure 6).

The "postal addresses" with which the hippocampus can retrieve memories correspond to the location and time information, the where and when of a remembered experience. The nerve cells responsible for location coding become active whenever we approach, either factually or even just in thought, the location associated with the respective memory. The place neurons give our lives a spatial dimension and make it possible for us to remember where events in our lives took place or might possibly take place, because they also encode future scenarios. That's why it's also helpful, when remembering an event or a thought, to visualize the place (and the time, see below) where the event was experienced or the thought was thought. In 2014, the Nobel Prize in Medicine was awarded for the discovery of place neurons because of their enormous importance for our self-understanding.

In the meantime, the corresponding hippocampal time neurons have also been identified.[10] They become active whenever we approach, either factually or even just in thought, the time of day at which a remembered event took place (so we feel appetite at certain times of day, even without actually being hungry yet). The hippocampal time neurons give our lives a temporal dimension, an understanding of when and in what order events in our lives took place, but also could take place, because even plans or thoughts concerning future events are encoded with the help of these time neurons.

It is now known that this ancient navigation system, with which all vertebrates orient themselves in time and space and how they know, for example, when and where a predator lurks or when and where something nutritious can be found, is used even more extensively in we humans.[11] Thus, in addition to the dimensions of space and time, this navigation system also encodes characteristics such as *large* and *small*, *thick* and *thin*, or *healthy* and *unhealthy*. Objects with similar properties have comparatively closer neuronal connections, which allows for rapid linking (association), a basic principle of human thinking. It allows us to generalize experiences and project them onto new events.[12] Even if we have never seen or experienced something before, certain characteristics of an object or situation, if they activate the corresponding feature neurons, can bring similar past experiences to mind, and that can be quite vital. Even if I may never have seen a particular snake before, its shape reminds me of other snakes, which signals to me that it might be poisonous.

However, this lifelong accumulation of "memory addresses" poses a problem. For although these space and time coordinates require far less storage space than the associated memory contents, they permanently reduce the frontal lobe battery's memory capacity if new storage spaces are not constantly created, and System II would no longer be fully functional. To prevent a decline in mental performance, nature has found a solution as amazing as it is efficient: the unique and lifelong growth of the dentate gyrus, the brain structure at the entrance to the hippocampus.[13] There, new place and time neurons can be created every day, which then fully recharge the frontal lobe battery and not only maintain its capacity, but even increase it.[14] But what can grow can also shrink—with fatal individual and social consequences.

CHAPTER 5

GROWTH AND SHRINKAGE OF THE FRONTAL LOBE BATTERY

The only constant in the universe is change.
—Heraclitus of Ephesus (535–475 BC)

The Lifelong Evolution of Individuality

Once again I wade in my rubber boots through the shallow riverbed, bow and arrow at the ready. But far and wide there are no ducks to be seen. Nevertheless I savor the water's quiet splashing, the current's gentle tug against my legs, and the forest's damp scent. I look up and admire the roof of branches and leaves that spans the stream and lets only a few rays of light through. An amazing construction. How do the trees that line the stream know where to grow, I wonder? How do they manage to form their leaves so that they interlock like the fingers of two hands? Not to draw water but gather as much sunlight as possible. One might think the trees have eyes with which they can see where there is still a patch of open sky to conquer.

It's a little miracle, I thought back then as a boy. And even today—despite or perhaps because of my studies in genetics—I am fascinated by this ability of deciduous trees to form their branches and leaves exactly where there is sunlight to harvest, and not to waste energy doing so in shady areas. But how does a tree know where to grow? It doesn't. Its shape develops through an evolutionary process of chance and necessity. Its genetic material ensures that new shoot axes and petioles are constantly forming—but completely independent of light and shadow, purely by chance. In the next step, the

39

tree's environment shapes its appearance through selection, or the preservation of those shoots and leaves that manage to capture the most energy-giving sunlight. The same applies to the associated further growth of the corresponding branches. They follow a necessity, because ultimately the entire tree's survival depends on the efficient capture of sunlight. On the other hand, leaves that the tree formed in places that do not receive light do not contribute to this necessity and therefore die. This evolutionary design principle of chance and necessity shapes not only trees but applies throughout nature, including the origin of species. The same principle that shapes the unique form of a tree also influences the development of our individual personality. In the brain, it is not branches and leaves but nerve cells (neurons) and their connections (synapses) that provide the unique form and function. These are formed on the aptly named dendrites, from the ancient Greek *dendrites* meaning "belonging to a tree," where they naturally capture not sunlight, but information instead.

In embryonic development and after birth, about twice as many neurons are created as are actually needed. Similar to the branches and leaves of a tree, only those neurons survive that are needed and contribute to adaptation and survival. Through an initial cellular surplus, it is ensured that sufficient neurons are available for all adaptations and thus all possible manifestations of the individual. The formation and strengthening of synapses also follows the principle of chance and necessity. This enables us to learn any language, any profession, and any faith. It doesn't matter whether one grew up in the Stone Age or today in the highly technological industrial age. The brain was and is prepared for everything.

Crucial for the development of our personality, however, is not factual knowledge, but above all our individual experiences, which survive particularly with the hippocampal neurons and cause their synapses to become stronger or weaker. The resulting unique combination of synaptic nerve cell connections within our entire neuronal network (new memories from the hippocampus restructure large parts of the rest of the brain during sleep) makes us who we are and dictates how we feel, think, and ultimately how we act. But what happens when we reach adulthood? Is further development then no longer possible? Is our personality fixed and unchangeable from this point on? According to the dogma of the Spanish Nobel laureate Ramón y Cajal, that would indeed be the case. In 1928, he described the fate of every nerve cell and thus also that of an adult's brain as follows: "Everything may die, nothing can be regenerated."[1] The end of any neurogenesis in the adult would not only bring about a developmental standstill, it would also mean that mental decline threatens with aging. That would be like a society in which there are no more births: In the long run, it would die out.

It is now known that the archicortex not only forms the only brain structures that can learn extremely quickly, but is also the only brain region that has the ability to form new nerve cells even into old age. The first evidence for this was already available in the mid-1960s, but the adult neurogenesis of the archicortex, the formation of new nerve cells in adulthood, was only very hesitantly accepted in the world of science. Despite clear evidence, overcoming the dogma of Ramón y Cajal was a difficult process that lasted until the beginning of the twenty-first century.[2] Yet, the first research results on the lifelong growth of the hippocampal memory store opened up a whole new perspective on the plasticity (malleability) of the adult brain and, not least, on the associated chance for further personal development of each individual into old age. This effect, among other things, could be shown in mice that were genetically identical. As part of the experiment, forty adult test mice were allowed to leave their dreary cage and spend several months on a large adventure playground enriched with many play opportunities. Although all had the same genetic makeup and cage history, they showed differences in exploring their playground over time. Initially small deviations in behavior resulted in increasingly more pronounced individual character traits. The mice that explored their surroundings most curiously also had the highest rate of adult hippocampal neurogenesis, which had a self-reinforcing effect and increased their curiosity and their thirst for knowledge even further.[3] The results of the study show that random differences in experience due to adult hippocampal neurogenesis lead to serious differences in psychological development in the long run. If, for example, a toy excites interest or curiosity, then curious behavior is promoted. Conversely, those who show rather passive behavior also develop no or less curiosity. Thus, chance and necessity—via neurogenesis—are at work in the formation of character traits.

Almost half a century after the American neurobiologist Joseph Altman discovered adult hippocampal neurogenesis in rats,[4] the long controversy over whether new nerve cells can also be generated in adult humans could finally be settled. In one study, an international team of researchers was able to show that the hippocampus still has a growth potential of several thousand new brain cells per day, even in adults.[5] And surprisingly, the potential for the hippocampal rate of new nerve cell formation is just as high in ninety-year-olds as it is in eighteen-year-olds. The ability for lifelong hippocampal neurogenesis is very likely fundamental for the evolution of wisdom in old age and ultimately even of human culture. Only through the continuous emergence of new memory cells is it possible to accumulate experiential knowledge over a long life and pass on the constantly expanding wealth of experience (without the risk of depression or Alzheimer's; more on that later) to the next generation. The evolutionary advantage of hippocampal neurogenesis, which is possible into old age, could be indirectly traced back and proven until the beginning of the twentieth century. Studies of marriage registers and church records showed that the presence of a grandmother in a family significantly increased the number of her grandchildren and their chance of childhood survival. Every decade that a grandmother lived after menopause gave her family, on average, two more grandchildren who reached adulthood, compared to families in which the grandmother died early.[6] Incidentally, a similar effect by grandfathers could not be proven, which could be due to the fact that fathers cannot usually be sure whether the offspring are actually theirs (for the longest time in human evolution, there were no paternity tests).[7] This could also explain why the maternal grandmother has a stronger effect on the blessing of children than the paternal grandmother. The longevity of the grandmother thus ensured that with each grandchild, more copies of her genetic material survived. Genetic information that positively influenced their longevity was thus also passed on much more frequently. Longevity and the ability to accumulate and pass on experiential knowledge even in old age by means of adult hippocampal neurogenesis are thus mutually reinforcing genetic selection criteria. The fact that women can fundamentally live long after menopause is therefore attributable not only to the achievements of modern medicine or to supposedly better living conditions. Rather, the predisposition for longevity (which is of course also passed on to the male members of a family) and the potential for mental fitness even into old age are due to an evolutionary selection process. This was already of great importance in the early history of mankind, especially but not only because there was no writing yet to preserve knowledge. Thus, longevity and the associated acquisition of individual experiential knowledge were a decisive advantage in the evolutionary struggle for survival in the Paleolithic Age. The earlier objection to this thesis, that hunter-gatherer cultures have an average life expectancy of only about thirty years and therefore such selection could not take place, has now been proven to be unfounded. This comparatively low average is based solely on a very high infant mortality rate. In hunter-gatherer cultures that still exist today, those who survive childhood have a more than 60 percent chance of reaching an age of over seventy years even without modern medical care; even encountering eighty-year-olds is no exception—and this despite the fact that the average life expectancy is still well below forty years due to the high infant mortality rate.[8]

Adult Neurogenesis Keeps You Young and Increases the Frontal Lobe Battery Capacity

When newly formed hippocampal neurons are experimentally destroyed in adult mice, it leads to an impairment of their spatial memory and the ability to remember and retrieve new content associated with it.[9] Thus, adult hippocampal neurogenesis is particularly necessary for gaining new life experiences. For this purpose, newly formed hippocampal neurons have the special ability to store individual events, even if they are very similar, as separate experiences.[10] To ensure this highly sensitive memory performance, they have very high structural and synaptic plasticity, which means that they can change quickly and efficiently in terms of both their networking and their transmission strength. This makes them as capable of learning as the nerve cells of a newborn.[11] This juvenile characteristic of being able to store new experiences so superbly distinguishes the *entire* hippocampus of young children, and you get to feel them when you, as an adult, try to beat them in a memory game. Even though this is practically impossible in most cases, the lifelong generation of new hippocampal neurons extends the developmental plasticity of childhood at least partially into adulthood and then into old age. Thus, adult hippocampal neurogenesis also ensures that the memory store for personal memories and thoughts in adults can maintain its youthful thirst for knowledge, as well as the ability to use System II in its everyday decisions.

Adult hippocampal neurogenesis takes place in the hippocampus's entrance area, the aforementioned gyrus dentatus. Since the place and time neurons already present there have the task, as "postal addresses," to make previous experiences retrievable in order to recall them into memory, a constant supply of fresh nerve cells is also needed there to become new place and time neurons for new memories. The generation of sufficient fresh nerve cells with youthful synapses is therefore essential so that new thoughts and experiences can continue to be stored and remembered without overwriting previous ones. This means that adult neurogenesis is necessary for hippocampal learning to take place efficiently (i.e., without loss of previous experiences) throughout life, thus making the wealth of experience on which our considered decisions are based ever more extensive.[12] If adult neurogenesis is disturbed, our spatial and temporal memory as well as our entire episodic memory suffer. In addition, the frontal lobe battery increasingly loses capacity. Without active adult hippocampal neurogenesis, our actions are dominated by stereotypical thinking (System I).

Adult hippocampal neurogenesis thus has two basic functions. Firstly, it enables lifelong, experience-based learning. Secondly, it maintains the frontal lobe battery capacity for the use of System II—and can even expand it throughout life. The limiting or storage-limiting factor for the hippocampus, which serves as the frontal lobe battery, is thus adult neurogenesis. That is a crucial insight. Creating the conditions for successful hippocampal "staying young" is therefore essential for our mental and psychological health—and how to achieve this is the subject of the following chapters.

What Can Grow Can Also Shrink

Although the frontal lobe battery's potential to maintain and even increase its capacity throughout life, as well as the strengthening of the frontal lobe's executive functions associated with it, is something we are born with, it ultimately depends on many different factors, as the following examples from the animal world illustrate.

Black-capped chickadees gather food in the summer and fall, storing supplies for the cold season. The high number of storage places, up to several thousand, is intended to ensure that not all of them will be plundered by spring, because otherwise these birds' lives would be endangered. However, they would face the same danger if they could not remember the many storage locations. For the small chickadee brain to perform this enormous feat of memory, it has a brain region comparable to the human hippocampus, which grows with each stored supply. However, the growth in autumn is followed by shrinkage in spring: When the food stores are no longer needed because enough fresh food can be found, then the corresponding memory stores also disappear.

There are several reasons for the shrinkage. The chickadees do not have to cover the increased energy demand of a larger brain all year round.[13] And without the shrinkage in spring, their brain would continue to grow year after year, which would pose an anatomical problem.

This adaptation of memory for such seasonal tasks has also been demonstrated in many other bird species. A corresponding size adjustment of the hippocampus was also observed in kangaroo rats when gathering winter stores.[14] It was always observed that experiences that are only important for a certain time make the brain region comparable to the hippocampus grow, but also shrink again when the specific knowledge is no longer needed.[15] In other animal species, further reasons for hippocampal size adjustment have been found. For example, the memory storage of voles increases by up to 23 percent during mating season, when new social bonds and tasks are pending, only to shrink again afterward. The growth but also the shrinking of nerve tissue is thus a completely natural process.

As mentioned at the beginning, a significant part of our individual development also consists of the fact that from the original surplus of neurons, those that were not able to establish functional contact points (synapses) with others disappear again. Without a synaptic "survival signal," they die. For the removal of unnecessary or, as shown in the animal examples, no-longer-needed neurons, these, like all other body cells, have a genetic "hara-kiri program." This is activated automatically as soon as the cell no longer receives life-sustaining signals.

Life-sustaining signals are transmitted in the case of nerve cells via synapses that they share with other neurons and are roughly as follows: "We need you!" So the synapses in the hippocampus don't just serve as memory storage and thus as a battery for the frontal lobe, they are also crucial for the survival of the communicating nerve cells and therefore for memories. However, this also means that the survival of new brain cells created through adult neurogenesis depends on us actually experiencing something "remarkable." Without a task, and in the case of the hippocampus this means without storing new experiences or thoughts, there is no purpose in life for newly formed nerve cells and thus no survival. This could be the reason why active adult hippocampal neurogenesis inspires natural (almost childlike) curiosity, because experiencing and storing new things prevents the cellular suicide of newly formed nerve cells. The cell's own suicide program is scientifically called apoptosis. In ancient Greek, *apoptosis* means "falling off." The process actually has some similarity to the falling of an autumn leaf. In a genetically orchestrated process, all nutrients are withdrawn from the leaves at the end of the year so that they are available for new leaves the following spring. In the apoptosis of nerve cells, their building materials are also returned to the rest of the organism through a kind of self-digestion in order to be able to use them again for new nerve cells. Basically, neurons kill themselves to protect all others or the rest of the organism through their altruistic sacrifice. If, for example, in the early development of a human, all the excess neurons produced were to stay alive, including those that are ultimately not needed, our brain would be far less efficient. It would also constantly require considerably more metabolic energy to keep these unemployed "freeloaders" alive, which could endanger the entire organism's life in times of scarcity. Their sacrifice thus serves the surviving cellular community.

Nerve cells are therefore "social beings" that communicate via their synapses, send and receive life-sustaining messages, or sacrifice themselves for the community. The latter also happens when they are attacked by viruses or bacteria, which lifts the blockade of the suicide program. Here the motto "one for all" applies because the infected neurons commit suicide to protect their neighboring cells and ultimately the entire organism from a spreading infection. Neurons also die, however, when they lose an already established contact with other neurons, their synapses become silent, and they no longer receive survival signals. This is the reason why, in Alzheimer's disease, the destruction of synapses leads to a very rapid degradation of brain tissue, particularly in the hippocampus area.[16] But nerve cells also die if memory content is no longer retrieved, perhaps because one is lonely and can no longer interact with others, or if they are simply no longer needed.

Unnatural Capacity Loss of the Frontal Lobe Battery

In contrast to the examples given from the animal kingdom, the experiential knowledge of humans is rarely only of seasonal or short-term importance. For a species whose survival is based on the lifelong maintenance of experiential knowledge and social bonds (which are also based on complex experiences), much of what is experienced, felt, and thought must also be retrievable as a memory even into old age. Since the accumulation and passing on of life experience were decisive criteria in the evolutionary struggle for survival of humans, the human hippocampus is genetically programmed for lifelong growth. Therefore, its degeneration, as occurs in Alzheimer's disease, among others, is unnatural and rightly considered pathological.

However, hippocampal growth is not automatic. For example, if we don't learn anything new, there is no need for new nerve cells and therefore no need for a growing memory storage or frontal lobe battery. Then adult neurogenesis is impaired, because ultimately, only those newly formed nerve cells survive, the synapses of which, as carriers of memories, are actually needed and used. This becomes a serious problem, for example, when older people become lonely, as is increasingly happening in our modern world due to the loss of extended families. Thus, the daily social contacts and tasks that could provide emotionally meaningful experiences or conversations, which would then be stored in the hippocampus, are lacking.[17] However, there are many other aspects of our modern lifestyle (which we will discuss in detail later) that lead to the hippocampus not growing throughout life, but shrinking a little every day and thus losing storage capacity. Thus, on average, the hippocampal volume of a person classified as "normally healthy" in studies decreases by about 0.8 percent per year of life.[18] According to a meta-study (which is a larger study that evaluates and summarizes several previous studies), the annual shrinkage rate in adults, also considered healthy according to the study, averages as much as 1.4 percent by volume.[19]

This enormous rate of regression of hippocampal volume in the general population was confirmed in a British study published in 2019 with data from almost twenty thousand Britons.[20] Again, the participants in the study were classified as mentally healthy, and the results obtained were declared to be normal values that can serve as a comparison to the values of actually pathological processes, such as depression or Alzheimer's disease, just as if the progressive hippocampus shrinking in the general population were a natural and not also a pathological process. Since in dementia, as well as in depression, a loss of hippocampal volume is not only part of the disease process but precedes it, this is to be regarded as a sick normality that is by no means natural. It is easy to explain, on the basis of this "normal" but completely unhealthy development, why, according to the World Health Organization (WHO), depression has become the number one widespread disease[21] and that dementias, particularly of the Alzheimer's type, are on the rise and now represent the third most frequent cause of death in the United States and Europe.[22]

The problem here is that in interpreting such hippocampal growth studies' results, no distinction is made between *normal* and *natural*. This is also true in everyday life, with fatal consequences. Because *normal*, in terms of the study, means nothing other than normality in our modern, technological everyday culture and lifestyle; how the majority does. However, this is now far removed from what would be natural for us humans due to our evolutionary development. We easily forget, do not remember, or simply do not know that our brain, although it has enormous potential and freedom to learn a wide variety of languages, technologies, and cultural behaviors, is not free from natural needs. Because of this ignorance, our brain is deprived of many factors essential for maintaining mental strength. But not only are there deficiencies, our brain—and especially our frontal lobe battery—is subject to many harmful influences, which will also be examined in the following. The fatal combination of too many deficiencies and too many toxins results in a chronic loss of capacity, with serious individual, social, and cultural consequences.

Our children also grow up in this unnatural environment. It must therefore be assumed that their frontal lobe battery generally does not develop its full growth potential from the outset, especially since the maturing brain is particularly sensitive to deficiencies in essential factors, to an increasing burden from toxins, and to academic and social stress. The results of studies show that every form of long-lasting stress has an inhibiting influence on the early growth of the child's hippocampus.[23] The increasing stress that parents are also under in our performance-oriented society does not go unnoticed by their children either. Thus, not only does hippocampal growth suffer in the parents, but the consequences can also be demonstrated in a hippocampal growth disorder in adolescents. Only a shift "Back to the Future"—toward a more species-appropriate life with a healthier diet and more exercise—offers natural protection.

Chronic Frontal Lobe Weakness

In addition to the growth of the frontal lobe battery, adult neurogenesis is also responsible for the development of a healthy curiosity and the building of high psychological resilience. Both are necessary for us to be open to new experiences—and thus they are key to building a rich store of experience that our frontal lobe can then draw on.

A disruption in the formation of new nerve cells in the hippocampus, however, has exactly the opposite effect. The consequence of a hippocampus that has not matured to its potential or is shrinking (usually both are the case) is a relative lack of perseverance and interest in pursuing new paths. US researchers argue in a review article that "damage to the hippocampus can lead to inflexible and non-adaptive behavior when high demands are placed on the generation, combination, and flexible use of information. This is reflected in such diverse abilities as memory function, navigation, curiosity, imagination, creativity, decision-making, assessment of characters, building and maintaining social bonds, empathy, social interaction, and even language use."[24] Furthermore, a growth disorder of the frontal lobe battery leads to a reduction in social intelligence, empathy, and even self-confidence.

This long list of negative effects of a growth disorder and a chronic loss of performance of the hippocampus is identical to that resulting from damage to the frontal lobe. If the same functional disorders occur in both cases, this can be considered further evidence that the hippocampus is indeed the frontal lobe battery. Thus, a chronic loss of frontal lobe battery capacity leads to equally chronic frontal lobe weakness. Because virtually no free synapses are available for storing new thoughts, it is logically very difficult for those affected to run through future scenarios in the frontal lobe using System II and to weigh things against each other, let alone to plan and implement them. Ego depletion becomes a permanent state from which one does not recover overnight.

Since the growth disorder of the hippocampus has become the unnatural norm due to our modern and largely species-inappropriate lifestyle and affects all generations, chronic frontal lobe weakness represents a massive individual (next chapter) and societal (last chapter) problem of pandemic proportions.

CHAPTER 6

CONSEQUENCES OF A SHRINKING FRONTAL LOBE BATTERY

Gutta cavat lapidem
(Constant dripping wears away the stone)
—Publius Ovidius Naso (43 BC–17 AD)

Everything Is Change

Eight o'clock in the morning. My grandparents are still asleep and I'm bored. As always in the mornings during my "Grandma and Grandpa" vacations, I sit in their small kitchen and wait impatiently for them to finally wake up. I'm looking forward to playing chess with my grandfather; it's the greatest thing for me. I glance at the kitchen clock, which doesn't have a second hand and therefore gives me the feeling that time is standing still. Therefore, I think I will have to wait forever until it is finally nine o'clock and my grandmother, usually the first of the two, gets up to make breakfast for everyone. The aroma of coffee will then lure my grandfather out of bed. I focus on the minute hand. One circle around, and boredom would finally be over. But it seems not to move at all. Odd. How can something change if no change can be perceived?

According to the findings of modern physics, things exist only if they interact with other things and are thereby also changed. The foundation of all being is thus constant change. Any form of standstill is therefore pure illusion.[1] This is physics, but also chemistry, and ultimately biology. It follows then that our hippocampus is also in constant flux. It has two fundamentally different paths it can take: It either grows or it shrinks.

The latter has become the unnatural norm. On average, the hippocampus of the adult population loses slightly more than 1 percent of its volume per year. That's about 0.004 percent of volume per day, which is hardly noticeable, but it still has consequences. For example, researchers at the New York Taub Institute observed a continuous loss of memory associated with the progressive reduction in size of the hippocampus.[2] This ongoing loss of personal memories is also a loss of what one could call the "self."

The Frontal Lobe Battery: More than Just Data Storage

However, the "normal," progressive hippocampal degeneration is only the tip of the iceberg of fatal consequences for those affected, because the hippocampus is far more than just a data storage device. It lies at the heart of our self-perception and thus of our psyche, our identity, our self-worth, our social and emotional intelligence, our ability to think using System II, and—not least because of its special role in stress regulation— also of our well-being.

Psychological Resilience

The hippocampus is a central regulator of our responses to stress. For this task, it fulfills all the requirements. On the one hand, the newly matured nerve cells that are integrated into the hippocampal network each day receive all the sensory impressions that signal a potentially stressful situation. At the same time, the birthplace of these nerve cells, the dentate gyrus in the entrance area of the hippocampus, is a repository of all "addresses" of the memory contents uploaded into the neocortex. The newly formed nerve cells in the dentate gyrus that are connected to the others thus have indirect access to all previous experiences that we have had in our lives. As a result, these new hippocampal nerve cells are ideally suited to assess every new life situation for its potential danger and to adjust our physical stress response accordingly. They do this by controlling the so-called stress axis or stress cascade, which consists of three sequentially connected hormone glands.

The first of these three is located in the diencephalon and is called the hypothalamus. This activates the second, the pituitary gland, via hormones, which then also controls the cortex of the adrenal glands with the help of hormones. The adrenal cortex, as the third gland, ultimately releases the stress hormone cortisol into the bloodstream. This mobilizes our energy reserves (blood sugar level rises), increases blood pressure and heart rate, and puts us in readiness to fight or flee.

However, an immediate response to a stress trigger is not always appropriate. Sometimes it may be exaggerated, sometimes perhaps too weak, so that either too much or too little cortisol is released to adequately meet the challenge. But the hippocampus is capable of learning; it only needs to be able to compare cause (stress trigger) and effect (stress reaction). To accomplish this, the hippocampal nerve cells have a sensor, the cortisol receptor. It measures the concentration of cortisol released into the bloodstream as well as the concentration of cortisol that arrives at the hippocampal nerve cells. Its concentration can then be compared with the nature of the stressor or the dangerous situation and, if necessary, adjusted. If, for example, based on previous experiences, the stress trigger is recognized as nonthreatening, the newly formed hippocampal neurons inhibit the stress response by acting directly at the beginning of the stress cascade, in the diencephalon. The cortisol level drops.[3] Due to the learning effect that takes place, the release of cortisol is also slowed down more quickly in similar dangerous situations in the future. If there are no previous experiences, the stress response remains active until the stress trigger is no longer present or is no longer perceived as threatening.

However, such learning effects can also have a detrimental effect, for example, if traumas experienced in childhood or experiences perceived as dangerous in the now adult trigger a completely exaggerated or unnecessarily prolonged stress reaction through corresponding triggers.[4]

Post-traumatic stress disorders also have such a hippocampal learning effect as a background. In these situations, seeking psychotherapy is recommended. The constantly elevated cortisol level prevents adult neurogenesis and even has a neurodegenerative (nerve cell–destroying) effect, thus shrinking the hippocampus, which is why this disease is always associated with a reduced volume of the frontal lobe battery and, moreover, its volume reduction (due to constantly elevated cortisol levels) further accelerates, which increases the shrinkage rate beyond the average.[5]

Natural Curiosity

The entire stress regulation system therefore depends on whether productive hippocampal neurogenesis takes place and continuously supplies the dentate gyrus with new nerve cells, or not. A high hippocampal nerve cell formation rate is thus causally related to our resilience to psychological stress and adaptation to new, potentially stressful situations.[6] A disturbed hippocampal neurogenesis, on the other hand, causes exactly the opposite: Even in situations that are not threatening according to objective criteria, the cortisol level remains elevated for longer than necessary. Because this stressful state is perceived as unpleasant, one avoids new things and natural curiosity suffers, reducing the chance of gaining new, potentially lifesaving experiences. A healthy, growing hippocampus makes it possible to give space to our natural interest in experiencing new things and thus expanding our wealth of experience.[7] Accordingly, the willingness to want to have new experiences is directly related to a growing hippocampus and, ultimately, to its volume as well.[8]

Self-Worth and Contentment

The results of the latest research show that the size of our hippocampus is also causally related to the level of our self-esteem.[9] Thus, our self-esteem increases with its growth, whereas it falls with its shrinkage. The direction its development takes then has an enormous psychological impact on our entire lives—how we feel, think, and interact with others, what we believe we can do, and what we actually do.[10] Even contentment is directly related to the size of the hippocampus, which is why its shrinkage brings with it a multitude of negative effects. These range from general dissatisfaction with one's own life and an increased risk, as a result, of resorting to drugs or violence,[11] of engaging in criminal behavior,[12] of developing eating disorders,[13] and even an increased likelihood of developing depression[14] or Alzheimer's disease.[15]

A major reason for these long-term and ultimately life-shortening consequences is the fact that hippocampal growth—as we have seen—is causally related to our mental resilience, the ability to cope well with stressful situations. Lack of self-worth, dissatisfaction with life, and chronic stress are ultimately fatal in the long run.

Creativity and Imagination

The frontal lobe is considered the seat of human imagination. But without the hippocampal ability to quickly and effectively store and retrieve new thoughts, it is simply impossible to develop creative thoughts and plans. Another important function of adult hippocampal neurogenesis is therefore its contribution to creativity and productivity. Conversely, people with impaired nerve cell formation and a shrinking hippocampus remain trapped in their habitual thought and behavior patterns. An extreme example of this was Henry Molaison, after he was deprived of both his hippocampi. But this is not an isolated case. Due to its high learning capacity (and the associated high energy demands), the hippocampus is a particularly sensitive brain structure. As a result, even medium-term oxygen deficiency, as is possible with birth complications, can lead to selective damage to the hippocampus. Children with such birth trauma can develop an IQ within the normal range because the frontal lobe and the working memory located there are usually not affected by oxygen deficiency occurring during birth.[16] Also these children learn to speak normally and are initially largely inconspicuous. Only their memory for personal experiences, including that for space and time, is impaired by the hippocampal damage. For example, they can't remember the way to school. And as was discovered in a larger study, the lack of hippocampal memory capacity is accompanied by reduced creativity—even with a completely intact frontal lobe.[17] It can therefore be assumed that *any* chronic degenerative damage to the hippocampus—not only the relatively rare one caused by a birth complication, but also the volume decrease considered "normal" in the adult population—results in reduced creativity.

Mental Flexibility

Reduced mental flexibility is the logical consequence of impaired adult hippocampal neurogenesis. With a weakening frontal lobe battery, we are ruled by System I, and thus remaining in stereotypical life patterns always takes precedence over System II thinking with the chance for self-reflection and change. Also, the "age-related stubbornness" often observed in modern societies can thus be explained by the fact that more and more hippocampal nerve tissue is lost over the years. Added to this, as discussed previously, is a reduced psychological resilience and thus a fear of any kind of newness. Therefore, in the case of deficient hippocampal neurogenesis, no willingness to change one's lifestyle in a health-promoting way, for example, can arise. In fact, a large portion of the supposedly healthy "general population" is incapable of this. According to a representative survey by DAK-Gesundheit (one of the largest health insurance providers in Germany), a majority of all Germans state that they are afraid of diseases that, in most cases, are caused by our modern way of life that is alien to our species and are therefore, in principle, avoidable. These include cancer (69 percent of respondents), Alzheimer's/dementia (49 percent of respondents), and stroke (45 percent of respondents).[18] Nevertheless, very few people are willing to change anything about their lifestyle. The fear of change seems to be greater than the fear of ultimately fatal diseases.

A serious example of the inability to make the necessary lifestyle changes is, in my opinion, type 2 diabetes, formerly known as "age-related diabetes," which has now become the most common form of diabetes even in children (previously, it was type 1 diabetes, an autoimmune disease). The cause of type 2 diabetes is a chronically excessive intake of carbohydrates combined with a lack of physical activity. However, the developmental path leading to this serious metabolic disease is not a one-way street, which until recently was the prevailing view. In 2018, it was shown that even severe type 2 diabetes, which is characterized by the fact that life-threatening blood sugar imbalances can no longer be avoided by oral medication but only by insulin injections, could be cured by a simple lifestyle change with fewer carbohydrates and more exercise.[19] In the long term, this would be the only lifesaving measure, as we now know.[20] Nevertheless, it is implemented by very few. There is a lack of mental flexibility to dare to try something new, and a lack of perseverance, for which active adult hippocampal neurogenesis is a prerequisite.

Healing is not exactly welcome. Even if patients were willing to defeat their diabetes with more exercise, obstacles are put in their way, because diabetes cure is rather unwelcome in Germany, as Prof. Stephan Martin, chief physician for diabetology and director of the West German Diabetes and Health Center, explains: "Programs are therefore not funded because health insurance companies would have to factor in corresponding losses. . . . And if a committed patient can currently end insulin therapy through lifestyle changes, he is punished for it because he then has to pay for his blood sugar test strips out of his own pocket. The data from the DiRECT study are convincing. But for Germany, the following holds: Healing is rather unwelcome!"[21] So you have to overcome two hurdles for your health: the shrinking hippocampus with all its consequences, and the lack of support from a monetarily oriented health care system.

A Depression of Pandemic Proportions

A disturbed regulation of our stress perception due to a chronically inhibited rate of new nerve cell formation in the hippocampus has also been identified as the decisive cause for the development of depression.[22] Therefore, the effect of all antidepressants on the market is based on a direct or indirect stimulation of hippocampal neurogenesis.[23] This also applies to extracts of St. John's wort, which intervenes via various mechanisms but has hardly any side effects, which is why I favor it to support treatment. A successful antidepressant therapy is therefore always accompanied by an increase in the volume of the hippocampus. However, hippocampus growth is usually impaired by a lifestyle that is not in line with our nature (more on that later), which is why the drug therapy approach usually only provides short-term relief and why depression often returns after the end of treatment. In fact, most of the antidepressants on the market have an effect hardly better than dummy drugs (placebos).[24] Only with a simultaneous change in lifestyle, which inhibits adult hippocampal neurogenesis, can antidepressant therapy be permanently successful. However, this is becoming increasingly difficult, as our culture and thus the lifestyle of most people is moving further and further away from a natural one. This causes new and increasingly serious deficiencies in many areas of life. The increasing impairment of hippocampal growth or hippocampal shrinkage is then a logical consequence and explains, among other things, the worldwide increase in depression.

According to the results of a meta-study, over 350 million people worldwide were affected by depression in 2014—and the trend is rapidly increasing.[25] This makes depression the most common mental illness in today's world population. For example, in the German population alone, the disease increased from 12.5 to 15.7 percent in just eight years (from 2009 to 2017).[26] Based on the results of studies by WHO, it can be assumed that the number of unreported cases is even higher here.[27] In the context of the COVID measures, the development toward depression has accelerated again for several reasons (less exercise, more fast food, fear of infection, etc.). A corresponding study from the US concludes "that the prevalence [roughly speaking, the frequency] of depressive symptoms in the US during the COVID-19 pandemic was more than three times higher than before the COVID-19 pandemic."[28] This, too, is an indication that depression is a consequence of our lifestyle.

Depression is characterized by sadness, loss of enjoyment of life, increased feelings of guilt with simultaneously reduced self-esteem, poor concentration, and, last but not least, by a reduced mental flexibility. Depressed people often feel disconnected from their own self and their past. In English scientific language, this feeling is called "derailment," which literally means derailment.

Self-perception is disturbed, which causes anxiety and thus stress in those affected, even without external stressors, and ultimately again indicates a hippocampal malfunction (hippocampus as the place of our self and its perception). The problem intensifies and often becomes permanent because the chronically elevated cortisol levels caused by anxiety and stress prevent hippocampal neurogenesis in the long term. Although cortisol has an acute anti-inflammatory effect by blocking the immune system, chronic activation of the stress cascade releases a variety of proinflammatory hormones (cytokines).[29] These, like the elevated cortisol, efficiently block hippocampal growth.

Cortisol inhibits cell proliferation. The stress hormone cortisol has the primary task of preparing our organism for either flight or fight. To provide the necessary energy for this, it shuts down all processes in our body that consume energy and are not needed in dangerous situations. This includes all growth as well as any healing and repair processes. This is unproblematic in acute stress because it only leads to a short-term stress response. But under chronic stress, the growth of the hippocampus suffers in the long term. And it gets even worse: Cortisol, in permanently elevated concentrations, as is often achieved in chronic stress, has a neurotoxic effect and thus even promotes the death of nerve cells. This neurodegeneration particularly affects the nerve cells of the hippocampus, as these are especially sensitive due to their high concentration of cortisol receptors.[30]

As a result of all this, a fatal vicious circle develops that is difficult to break: A nongrowing, ultimately shrinking hippocampus further lowers stress resistance, which exacerbates the depressive mood and leads to a constantly active, no longer dampened stress cascade. The constantly elevated cortisol has a neurotoxic effect and accelerates the shrinking process of the hippocampus.

Pandemic of Chronic Fatigue

The same vicious circle of cause and effect is very likely also the basis of chronic fatigue syndrome (CFS).[31] Hallmarks of CFS are persistent fatigue, exhaustion, and lack of motivation. As with depression, a disturbed adult neurogenesis and hippocampal degeneration are at the center of events here as well.[32] In addition to the massive exhaustion that lasts for months, there is also a disturbed episodic memory and hypersensitivity to stress triggers, i.e. poor mental resilience. Many lifestyle factors contribute to this, such as a simple lack of micronutrients, which would be easy to fix (more on that later). Common to all these factors is that they are essential for hippocampal growth and every single deficiency inhibits hippocampal neurogenesis. Clinical studies show, for example, that a balanced diet with whole grains, polyphenol-rich (colorful) fruits and vegetables, and aquatic omega-3 fatty acids improves fatigue symptoms caused by ill-ness.[33] A micronutrient-rich, anti-inflammatory, and stimulating diet thus goes a long way in staving off fatigue symptoms.

Prevalence is the proportion of people affected by a particular disease at a par-ticular time or in a particular period of time compared to all those examined. In line with an increasingly unhealthy lifestyle, it is not surprising that the prevalence of CFS has increased massively since it was defined as an independent disease in 1994. Thus, about 1 percent of the world's population now suffers from it.[34] The *feeling* of being chronically exhausted is significantly higher at over 3 percent, as was found in large international studies.[35] Due to the fact that CFS is attributable to deficient hippocam-pal neurogenesis and degeneration of the hippocampus, it could also be explained as a result of a chronically weakened frontal lobe battery. Thus, one would have a causal therapy option along with the cause for this ominous disease.

Long COVID—A Hippocampal Growth Disorder?

Long COVID is a collective term for a large number of symptoms following a recov-ered COVID-19 infection, whereby constant fatigue and chronic lack of motivation as well as cognitive dysfunction are in the foreground.[36] However, this does not occur unexpectedly. A chronic fatigue illness (CFI) is a typical accompanying symptom of autoimmune diseases and cancer as well as the main symptom of the chronic fatigue syndrome described above. Some experts suspect that CFS, CFI, and thus also Long COVID represent essentially the same clinical picture.[37] The change in the senses of smell and taste could have a common root in disrupted adult neurogenesis, since this affects the hippocampus, but also the olfactory brain (rhinencephalon) as part of the archicortex at the same time.[38]

A micronutrient deficiency, such as a vitamin D deficiency, is usually found as the cause, which is further aggravated by the increased metabolism of the immune system due to the corresponding disease processes.[39] However, as we will see in more detail, almost all micronutrients are essential for productive adult hippocampal and rhinen-cephalic neurogenesis, and any deficiency inhibits this growth process. This problem—that is, adult neurogenesis disrupted by a micronutrient deficiency—thus explains the majority of CFS and Long COVID symptoms, especially since this deficiency is also the main cause of severe COVID-19 disease courses.[40] It is therefore to be expected that the entire archicortex is also affected in the case of this infectious disease. Fortunately, severe

COVID-19 courses can be largely prevented by correcting such deficiencies, which should also significantly reduce the possibility of developing Long COVID.[41]

Alzheimer's and Dementia—A Pandemic of Self-Destruction

Due to the chronic atrophy of the hippocampus, the risk of developing chronic depression or CFS increases, and also of developing Alzheimer's dementia, which is often associated with it. Parallel to the increase in depression and CFS, the number of people with Alzheimer's worldwide has more than doubled between 1990 and 2016, rising to almost 44 million—a trend that is dramatically increasing.[42] And we have seen that in the United States, Alzheimer's is now the seventh most common cause of death, and fifth over the age of sixty-five.

An early indication of impending Alzheimer's dementia is a chronically disturbed hippocampal regulation of the stress cascade.[43] Therefore, depression usually develops *before* the mental decline due to Alzheimer's—whereby again, at least in the initial stage, the stress caused by the noticeable decrease in mental performance can contribute to or worsen depression. According to the results of a Swedish long-term study, depression is thus not only a frequent consequence of Alzheimer's dementia, but usually also its precursor.[44] A comparable correlation with other forms of dementia, such as the second most common after Alzheimer's, vascular dementia (based on the blood vessels or arteriosclerosis), could not be proven. It is therefore possible and very likely that both diseases, depression and Alzheimer's, have a common cause. The most likely candidate for this is considered to be impaired adult hippocampal neurogenesis, as I have explained in a scientific review article.[45] Thus, in patients with depression, but also in the early stages of Alzheimer's, a rapid decrease in the proliferation of nerve cells in the hippocampus can be demonstrated (which is still quite possible in mentally healthy ninety- and even hundred-year-olds).[46] This is why the early symptoms of the two diseases are hardly distinguishable from each other.[47]

The fact that the disease process begins precisely in this part of the brain also speaks for a hippocampal cause of Alzheimer's dementia, and accordingly, the degree of hippocampal atrophy is one of the best markers for how far one has already progressed on the path toward Alzheimer's.[48] The best protection against Alzheimer's (and depression) would therefore be to keep adult hippocampal neurogenesis active throughout life or to reactivate it through an appropriate lifestyle.

Cognitive reserve. Education in one's younger years is said to protect against Alzheimer's or at least delay the disease process by a few years. This is attributed to the so-called "cognitive reserve," which is built up through education. Until now, however, it wasn't entirely clear what this reserve actually consists of. Now it appears that people with higher education have a larger hippocampus. Although it also shrinks in "educated individuals" at about 1 percent per year, no slower than in the general population, the larger initial volume explains the later onset of Alzheimer's dementia.[49] The cause of the degradation of hippocampal tissue thus lies in other areas of life and is not influenced by education.

Trapped in a Vicious Circle of Cognitive Decline

The pharmaceutical industry is always searching for remedies through medication to get a grip on the symptoms and long-term consequences of such "hippocampal" diseases (depression, CFS, Long COVID, and Alzheimer's). However, the actual causes are not eliminated by this. Perhaps this is not even desired, however, because after all, the business model is based on the sale of their products, as we already know from studies on type 2 diabetes—here, too, healing in the sense of a causal therapy, as discussed, is rather undesirable.[50] This is problematic, because only by eliminating the causes of a disease can its development be avoided or people still be healed if the corrective measures are implemented in time.

However, even doctors and therapists who are open to eliminating causes and want to convince their patients to change their lifestyle face a huge problem. Because the frontal lobe battery doesn't grow but shrinks in those affected, they show almost no interest in such radical measures. This is usually not even possible preventively, because adult hippocampal neurogenesis is usually already blocked before serious symptoms appear. Once the disease has broken out, the willingness to undergo curative therapy usually decreases completely. Thus, the patients, but ultimately almost all of humanity, find themselves in a deadly vicious circle. Finally, as mentioned, the hippocampus shrinks on average in all adults due to impaired neurogenesis.

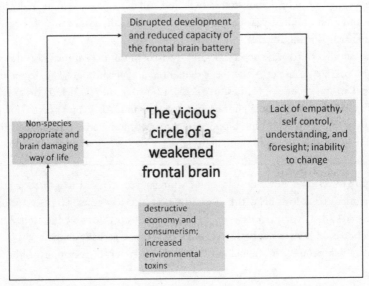

Figure 8: A deadly vicious circle.

All modern diseases of civilization ultimately have mostly the same cause, namely a species-inappropriate lifestyle, which causes the shrinking of the frontal lobe battery. Furthermore, the consequences of many of these diseases of civilization also hinder adult neurogenesis and promote hippocampal neurodegeneration. This includes, for example, chronic inflammatory reactions as a result of chronically elevated blood sugar in diabetics, or a reduced energy supply to the hippocampus due to arteriosclerotic damage to the blood vessels in people with high blood pressure. Breaking these vicious circles of chronic frontal lobe weakness is very difficult and ultimately only possible if those affected start early enough and are supported by a strong team of therapists and, ideally, by a life partner.

Consequences of Impaired Hippocampal Growth for Child Development

The hippocampus thus plays a crucial role in the development of personality and individuality, a healthy sense of self-worth, imagination, and creativity of psychological resilience, impulse control, perseverance, and, last but not least, empathy and a social conscience. All these fundamentally human traits depend on hippocampal neurogenesis functioning correctly. Since, as illustrated in Figure 9, the growth of the hippocampus is particularly rapid in early childhood, this phase of human development is particularly formative.

Meanwhile, however, more and more children are suffering from diseases of civilization such as adult-onset diabetes, high blood pressure, obesity, and arteriosclerosis, diseases that for a long time seemed to be mainly reserved for adults. Because these are based on the same causes that also inhibit hippocampal growth in adults, it must be assumed that it is already impaired in most children. This has the same detrimental consequences as in adults, but because certain phases of child development are formative, also some that are specific to children and adolescents.

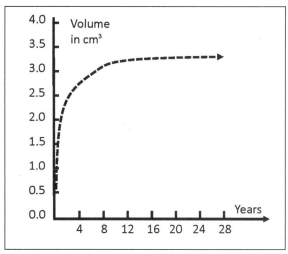

Figure 9: Growth of the human hippocampus.

Childhood and Adolescent Depression

More and more children and adolescents are suffering from depression and burnout. Burnout is just another term for the same phenomenon when the trigger is found in the workplace or, for children and adolescents, in school. According to the results of a study by the *Kaufmännische Krankenkasse Halle* (KKH) from 2017, depressive states in thirteen- to eighteen-year-olds increased by over 90 percent in the previous ten years. In total, it is estimated that approximately 1.1 million of six- to eighteen-year-olds in Germany suffer from depressive reactions to physical and mental stress caused by increased pressure to perform and bullying at school.[51] And according to a representative study by the DAK, almost every second student now suffers from chronic stress—with significant negative effects on health.[52] An overload of too much material in too short a time is certainly a relevant factor in the development of school stress. But what about the general resilience of children and adolescents and their ability to successfully cope with stressful situations? Curiously, weakened performance and reduced psychological resilience due to hippocampal growth disorder is not discussed as a possible decisive factor. Perhaps because in this way one does not have to question either the unnatural school system or the unnatural development of our children and thus our way of life as a whole. Even though a great many jobs depend on this unnatural way of life, a discussion of its consequences is urgently needed. Ultimately, our children suffer doubly from the excessive social and economic demands: on the one hand, from the growing pressure to perform, and on the other hand, from a lack of psychological resilience to deal with this pressure.

Decline in Empathy

The hippocampus is the gateway to our social memory and thus also to our social conscience. For example, the ability to empathize with other people, or living beings in general, depends on our emotional memories. A reduced capacity for empathy would therefore be a logical consequence of chronic frontal lobe burnout due to a low-capacity frontal lobe battery. If this is true, the capacity for empathy would have to decrease to roughly the same extent that depression and other hippocampal diseases are increasing worldwide.

Mark H. Davis, professor of psychology at Eckerd College in Florida, developed the Interpersonal Reactivity Index (IRI) to measure empathy.[53] The IRI consists of a catalog of twenty-eight statements that are rated by test subjects with a maximum of five points, using the statements ranging from "Does not describe me well" to "Describes me very well." Among other things, it examines "perspective-taking," that is, the ability to put oneself in the shoes of other people. Under "empathic concern," the IRI measures the extent to which respondents are capable of developing compassion for people who are in an unfortunate situation. A total of 72 IRI studies tested American college students over a period of thirty years (1979 to 2009). The young representatives of an industrialized society of the global north showed a frightening loss of empathy. As Figure 10 shows, both empathic concern and perspective-taking declined significantly in the groups studied. People with low scores in these two areas often behave in an antisocial manner.

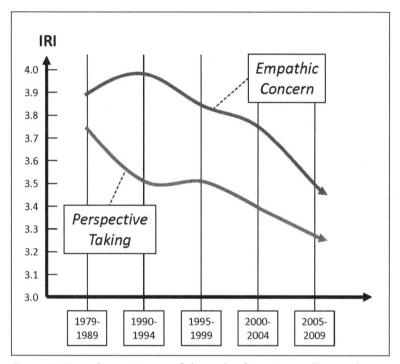

Figure 10: Empathy continues to fade: study of American college students.

Feeling empathy is a crucial factor when it comes to not exhibiting behavior that is harmful to other people. Accordingly, science also attributes the increasing destruction of the biosphere and biodiversity to a growing lack of empathy in our modern society. Gerardo Ceballos, biologist, ecologist, and conservationist at the National Autonomous University of Mexico and lead author of a study on species extinction,[54] told *USA Today*: "The massive loss of populations and species reflects our lack of empathy to all the wild species that have been our companions since our origins."[55]

However, it is not only empathic, social, and mental performance that suffers from impaired hippocampal neurogenesis. In extreme cases, it can even lead to clinically significant disorders that are specific to childhood and adolescence. It is obvious that their increasing occurrence is also due to an increasingly unnatural way of life that our modern society imposes on children. To make matters worse, in the young, developing brain—in contrast to the adult, largely mature brain—it is not only the natural development of the frontal lobe battery that suffers from many deficiencies and the increasing influence of toxic damage, but also the development of the frontal lobe itself. The resulting diseases are therefore usually a combination of developmental disorders or damage to both parts of the brain, the frontal lobe and its battery.

Schizophrenia

A good example of this problem is schizophrenia, which has long since ceased to be a rare disease. In Germany alone, there are almost one million people with the disease. More and more young people suffer from a disruption in the development of their individuality, which translates to "indivisibility." Conversely, the term *schizophrenia* is derived from the corresponding "division" of the soul; it is a neologism from the ancient Greek *schizein*, "to split or shatter," and *phren*, "mind or soul."

The division of personality due to impaired brain development is characterized by delusions, hallucinations, and cognitive deficits. Schizophrenia also includes hippocampal memory disorders; language that follows no logic, which makes it almost impossible for outsiders to follow the train of thought; and correspondingly confused behavior, as well as the inability to express emotions. A lack of motivation, interest, and empathy are also symptoms of this broad spectrum of illnesses. These functional problems are based on structural developmental disorders that can be detected as early as the early stages of the adolescent brain.[56] A typical characteristic is a reduction in hippocampal volume.[57] Accordingly, it has been shown that functioning hippocampal neurogenesis offers protection against the development of schizophrenia.[58] The importance of a growing hippocampus as a protective mechanism could lie in the fact that it continuously gathers life experience, on the basis of which our self-understanding and an individual personality can develop and also consolidate.

In the frontal lobe, a kind of "rewiring" takes place with the onset of puberty until the end of adolescence, until the age of about twenty-four. This results from the resolution of existing and the formation of new synaptic connections.[59] This process, it is suspected, helps to question acquired social dogmas in order to possibly advance the development of the group in which one grows up. This could have been an evolutionary advantage in the struggle for survival, which is why this pubertal mechanism became a significant part of our genetic program. This highly vulnerable and often conflict-laden time is therefore still an important process today in the maturation into an independently thinking and acting adult and crucial for the development of the individual personality.

Toxins such as nicotine, alcohol, and many others impair neurogenesis in the development and later also in the adult brain. If they are administered during the pubertal reorganization phase of the frontal lobe, they become causal risk factors for schizophrenia.

Traumatic experiences can also have a disruptive influence. Serious deficiencies in micronutrients, which are needed as building blocks and messenger substances, are also highly problematic in this phase, which is highly sensitive for self-development. For example, it has been shown that fish consumption reduces the risk of developing schizophrenia.[60] Fish provides many micronutrients, almost all of which can be considered responsible for this protection: Vitamin A, vitamin D, almost all vitamin B variants, some essential trace elements, and last but not least, the aquatic omega-3 fatty acids. These include EPA (eicosapentaenoic acid) and DHA (docosahexaenoic acid), which we obtain almost exclusively from fish and seafood. They are essential for us humans, because we can hardly convert omega-3 fatty acids from plant (terrestrial) sources, such as linseed oil, into EPA and DHA. I will discuss this further below.

The aquatic omega-3 fatty acid DHA is an essential building block for our brain. Furthermore, it is the precursor to a whole series of neuronal growth factors and thus has a special role in brain maturation and the lifelong increase in frontal lobe battery capacity. It has been shown that targeted dietary supplementation with these fatty acids can prevent the typical progressive loss of brain mass in schizophrenia patients.[61] Clinical studies have also found that aquatic omega-3 fatty acids attenuate psychotic symptoms—and this without the undesirable side effects known for all standard medications.[62] These new findings indicate that, in addition to the influence of toxins, which unfortunately are the norm in our modern society, the lack of such essential fatty acids could also be a decisive cause of schizophrenia. This conclusion is supported by another, very revealing study: thirteen- to twenty-five-year-olds who had already experienced a first episode of schizophrenia received either aquatic omega-3 fatty acids or a placebo for a short period of only three months. In this study, the likelihood of the omega-3 group experiencing another psychotic episode within the following nine months was reduced by almost six times compared to the placebo group. After another six years, 40 percent of the placebo-treated subjects developed full-blown schizophrenia, while in the omega-3 group, it was still less than 10 percent.[63] This result was as spectacular as it was unexpected, as the three-month "therapy" was several years in the past. Apparently, even the short-term correction of a deficiency of these essential brain building blocks in this period of psychological development has long-term effects.[64] The result makes it clear that in the pubertal and postpubertal phase up to young adulthood, aquatic omega-3 fatty acids must not be missing from the diet in order to develop one's own personality that questions social norms. Thus, developmental disorders with functional weaknesses of the frontal lobe and battery could be prevented, as could, in extreme cases, even serious diseases. Adequate supply of aquatic omega-3 fatty acids and many other nutrients essential for brain development, such as vitamin D, would increase the chance of healthy brain maturation in every child, as would a more general species-appropriate way of life. Unfortunately, however, all of this is the exception and a chronic deficiency, even in childhood, is the rule.

Attention Deficit Hyperactivity Disorder (ADHD)

In more and more children, the development of the frontal lobe and its battery is slowing down due to such deficits in our modern diet. The widespread lack of sleep and exercise also hinders natural brain development. The disruption of frontal lobe maturation and the slowing of hippocampal growth caused by such factors (and there are several more that we will discuss) delay school readiness and ultimately also prevent a level of schooling that would potentially be achievable for the child.

At the same time, despite this fatal development in our increasingly performance-oriented society, the demands on children are rising, as is the pressure of expectations from parents. Thus, the age at which a child is considered ready for school is also being pushed further and further forward—after all, today's children are supposed to enter working life earlier and earlier and contribute to economic growth. Thus, society demands that they adapt to the lifestyle of the modern adult, such as getting up early and sitting still for long periods. However, this only works with an already advanced frontal lobe maturity—even if it is still unhealthy then. Fewer and fewer children can withstand this trend toward an unnatural, economy-conforming way of life. This cultural development, which on the one hand slows down brain maturation, but at the same time also requires accelerated brain maturation, could be an explanation for the absolute as well as relative Attention Deficit Disorder (ADD) of more and more children. Sometimes the ADD is combined with a hyperactivity disorder (H), that is, a lack of self-control or impulse control. This overall picture is then referred to as Attention Deficit Hyperactivity Disorder or ADHD.

Time and again, ADHD, now one of the most common childhood diseases, is declared to be linked to a genetic predisposition.[65] This is somewhat reassuring for parents because, after all, they haven't done anything wrong. But it is also for the treating physicians, because if the genes have a malfunction, life-changing measures do not help, only medication, and that is easy to prescribe. However, Ritalin and co. only mask the symptoms of the developmental delay, while the actual causes are neither eliminated nor addressed in practice. This threatens the danger that the developmental delay will become a chronic developmental disorder that can last into adulthood—which now actually happens in more than half of the cases.[66] The enormous frequency of ADHD, now occurring in more than 5 percent of all children, clearly speaks *against* a genetic cause, especially since the number of unreported cases is very likely even higher. For even if there were a change in the genetic makeup, a mutation (none has been found so far), it could not spread so quickly in the population that such enormous growth rates could be explained. That is the nature of genetic changes: They need many generations, many decades, to prevail under selection pressure. What can spread much faster than genetic changes, however, are cultural behaviors.

Common side effects of ADHD as a result of delayed development of the frontal lobe and its battery include:

- *Depression.* This occurs five times more frequently in young ADHD patients than in children and adolescents without this diagnosis.[67] It is usually considered an emotional, stress-related consequence of the ADHD problem—after all, a quarter of children diagnosed with ADHD suffer from (fear of failure) anxiety.[68] Chronic stress and anxiety are known triggers of depression. Another, at least equally plausible cause, and possibly even the primary one, could be a growth disorder of the frontal lobe battery, with a correspondingly reduced psychological resilience in the affected children.
- *Sleep disorders.* These are frequent and exacerbate the overall problem, because sleep is of enormous importance for the maturation and function of the frontal lobe, but as discussed, also for the maturation and function of its battery. Interestingly, a study was able to achieve an improvement in sleep quality simply by correcting a deficient supply of aquatic omega-3 fatty acids.[69] In fact, children with ADHD consistently have an insufficient omega-3 fatty acid status.[70] But there are far more causes than omega-3 deficiency that rob children of sleep (more on that later).
- *Reading and spelling difficulties.* These are found in almost half of the children with ADHD.[71] Interestingly, in children diagnosed with ADHD who, compared to their peers, were about two years behind in the development of their frontal lobe, a four-month intake of aquatic omega-3 fatty acids improved their reading and writing.[72] This, too, points to a nutrition-related deficiency of essential brain building blocks as a cause of ADHD.
- *Poorly developed working memory.* This leads in particular to problems with verbal instructions.[73] On the other hand, behaviors necessary for play are largely remembered, understood, and implemented without any problems—typically childlike, one might almost say.

Because a species-inappropriate diet with corresponding deficits and many other aspects of the economy-compliant development of our children are responsible for the absolute delay in the maturation of the frontal lobe and its battery, therapeutic or, even better, comprehensive preventive measures should start precisely there. For example, there is a correlation between ADHD risk and the frequency with which junk food is consumed, whereas a balanced and healthy diet represents a protective factor.[74] This includes, among other things, an adequate supply of vitamin D, the concentration of which in the blood is generally far below the recommended guideline values in children in Germany.[75] Correcting a deficiency in aquatic omega-3 fatty acids also improved the ability for positive social interactions and concentration, as a whole series of clinical studies showed.[76] Ultimately, all possible vitamin and trace element deficiencies should be addressed at all ages. These deficiencies can be diagnostically ruled out or corrected in children, ideally even in expectant mothers.

Autism—Permanent Immaturity

Autism encompasses another spectrum of neurological developmental disorders of the frontal lobe. They are characterized by impairments in social interaction and communication, as well as by repetitive and stereotypical behavior patterns. Due to a view reduced to these aspects, Asperger's syndrome is also included in the autism spectrum. I consider this to be a serious mistake, because there are significant differences. For example, children with Asperger's syndrome, in contrast to those with so-called high-functioning autism, do not show any delay in language development. On the contrary—they typically start speaking early, often have special talents in rational-analytical thinking, and are usually of above average intelligence. It may therefore well be that this is not a delayed development, but only a different one. Asperger's would therefore not be a disease, but a blessing for humanity, since people with Asperger's, with their ability to perceive the world somewhat differently, are responsible for many scientific and technological advances.[77] From a sociocultural perspective, Asperger's syndrome is thus an invaluably important part of the human developmental spectrum.

If Asperger's is excluded, those affected by actual autism are usually significantly reduced in intelligence. Further core symptoms include ADHD, anxiety disorders, and sleep problems. This form of autism is to be considered a disorder of frontal lobe development, which usually becomes recognizable by the age of about three years at the latest. The fact that the cause of the frontal lobe developmental disorder is very likely to be found in cultural maldevelopments can be seen in the increased incidence of the disease. In the middle of the last century, only one in about two thousand children suffered from autism, which corresponds to a prevalence rate of 0.05 percent. The incidence of the disease is now over 2 percent, which is more than forty times higher.[78] No other disease of the child's mind, not even ADHD, has increased so rapidly.

This rapid increase in the frequency of autistic disorders in our society occurred parallel to the increased consumption of foods rich in omega-6 fatty acids (sausages and meat products, omega-6-rich oils) and at the same time low in aquatic omega-3 fatty acids.[79] The absolute deficiency of the aquatic omega-3 fatty acid DHA in the child growing in the womb is exacerbated by the excess of the omega-6 fatty acid arachidonic acid (AA) in the diet of the expectant mother, as she passes it directly to the fetus. The two fatty acid classes block each other, both during transport across the placenta and afterward during transport to the brain. A relative excess of AA thus inhibits the supply of DHA, which is already less available from the outset, to the fetal brain. Overall, this leads to disrupted synapse formation throughout the brain, but especially in the rapidly growing frontal lobe during the last trimester of pregnancy, and thus to an impairment of the networking of its nerve cells, the known disruptions that are characteristic of autism.[80]

In contrast to schizophrenia, correction outside of this early childhood timeframe is difficult in autism. Nevertheless, an initial study shows that treating autistic children with aquatic omega-3 fatty acids has a beneficial effect on concentration, eye contact, language development, and motor skills.[81] Ultimately, however, only the prevention of such serious disorders is promising, which makes it absolutely necessary—also because it must be assumed that the diagnosis of autism is very likely only the infamous tip of the iceberg. Hardly any fetus or toddler receives all essential micronutrients in sufficient quantities to be a species-appropriate diet.

Drug Addiction

Drug addiction is both a disease in itself and a symptom of many mental disorders. Often, however, drugs are also their cause. Experimentation with drugs and the slide into addiction typically happen in adolescence. Puberty, in particular, with the restructuring of the frontal lobe, is an especially vulnerable phase for this. It appears that a disrupted interaction between the frontal lobe and the hippocampus is of particular importance in the development of addiction problems, and impaired hippocampal neurogenesis is therefore considered a cause of drug abuse.[82] This is due to the already well-known long list of psychological problems caused by inadequate nerve cell formation in the hippocampus, such as low self-esteem and increased sensitivity to stress, with a tendency toward depressive moods, which impair natural self-protection.

Accordingly, hippocampal growth disorder increases both the susceptibility to addiction and its severity. In addition, most addictive substances such as nicotine, alcohol, opioids, and cocaine further suppress hippocampal neurogenesis, thus increasing the urge to use drugs and psychological dependence.[83] Conversely, however, and this can be seen as another indication of hippocampal involvement in drug addiction, increasing neurogenic activity is an important and ultimately very likely even the best therapeutic strategy for dependencies.[84]

Lifelong Mental Health

We wish only the best for our children and grandchildren: that they grow up happy and learn to master their lives with curiosity, imagination, and a healthy self-confidence. At the same time, we ourselves hope for a long life, so that with full mental fitness we can also support our grandchildren in their development with advice and action. These hopes and desires are very likely part of our genetic program—the "evolution of the grandmother" is the best indication of this. Lifelong mental health is instilled in us from birth; it is based on the potential for lifelong growth of the frontal lobe battery.

The effects of a disturbed development of the frontal lobe battery and its chronic loss of capacity, which has increasingly become the unnatural norm, are all the more dramatic. The reason for this is the growing gap between what the human brain needs for development and maintenance of its functionality and how we shape our modern world. Ultimately, it is the seemingly unbridgeable discrepancy between an economically compliant and a species-appropriate way of life. But subordinating the entire cultural development of humanity to purely economic interests is based on the historical error that man stands above nature. This is wrong, because a species-appropriate life outside nature is completely impossible. We must correct this enormous cultural-historical error if we want to lead a life in harmony with our own nature and thus increase the chance that our most elementary wishes and hopes will be fulfilled: Long-term health and mental fitness for the whole family.

CHAPTER 7

EVOLUTION OF
THE HUMAN MIND

If a plant needs twelve substances to grow, it will never grow if even one of these substances is missing, and it will always grow poorly if even one of these substances is not present in the amount that nature of the plant requires.
—Carl Sprengel (1787–1859)

Laws of the Minimum and the Maximum

I feel like an intensive care patient. However, I'm not lying in a bed, but sitting on a racing bike. It is hot, it would be over 113 degrees in the shade, but there is no such thing. The sun is blazing from above, and the asphalt is shimmering with heat below. I have just reached the Mojave Desert, notorious for its relentless heat, a few hours after taking off from the US southwest coast, followed by my medical monitoring team in an air-conditioned car a few meters behind me. At regular intervals, a voice over the loudspeakers urges me to drink a salty nutrient solution. If I don't leak within the next half hour, my urine could not be tested for mineral content and density. Moreover, this would be a clear sign that I had not drunk enough. Then I would have to interrupt the bike race until I was sufficiently hydrated again. But after just over 220 kilometers, there are still over 4,600 kilometers ahead of me, and I have just over eleven days left to successfully finish the Race Across America (RAAM), the world's toughest endurance race. So I drink, even if it exhausts me and I don't feel thirsty.

Hopefully, nothing goes wrong. After all, in order to reach the goal on the east coast of the US, I not only trained a lot, but also mobilized a large team of supervisors for the joint adventure. To increase my chances of success, I've delved into the missteps of those who came before me, studying their every wrong turn and stumble along the way. Even very experienced professional racers had failed to meet the challenge despite the most intensive preparation.

There were plenty of reasons for that. Lack of fluids is one of them. But also, a lack of sleep. In this nonstop race from coast to coast, everyone tried to get by with just under one to two hours a day, according to the motto "He who sleeps loses." However, lack of sleep leads to hippocampal information overload and, inevitably, to "uploading" while awake: Paranoia is inevitable! Microsleeps, especially on long descents, have also sent many RAAM riders to the hospital, and some even to their graves. But it's not just the hippocampus that needs regenerative breaks; the muscles, tendons and ligaments also need them. However, with this unnatural racing strategy without sufficient recovery, inflammation occurs, which can lead to premature termination of the race. Deficiencies in nutrition are another cause of failure. Ten thousand calories and up to twenty liters of fluid are needed daily, so you can go wrong in a lot of ways Too little of one or the other vitamin or mineral, and the race is over. With every report I read in the lead-up to the race, the checklist for my "intensive care team" grows longer.

In 1838, the agricultural scientist Carl Sprengel, quoted at the beginning, also wrote a checklist. This was based on a fundamental understanding of the optimal growth of plants. The law of the minimum explained what a plant must not lack to thrive and bear many fruits, thus increasing its chance of reproduction and evolutionary success. This is measured by the number of reproductive offspring—in the case of plants these are their fruits—and thus Sprengel's law of minimum describes the result of a species-specific evolutionary adaptation. For example, a cactus may like it sunny and warm, while a fern prefers it shady and cool. Why is that? The reason for this lies in their species-specific evolutionary history.

In order to survive and reproduce in the desert, cacti acquired the ability to endure drought and great heat. They have few or no leaves to keep their surface as small as possible and, at the same time, store a lot of water. On the other hand, in order to thrive and spread on the ground of dark forests, ferns learned to get by with little sunlight. For this purpose, they have very broad, mostly rosette-like arranged leaves with a lot of chlorophyll (light collectors). But despite these enormous differences, there are at least three important similarities:

1. Both plant species adapted to the respective living conditions that prevailed during their long evolution.
2. Both still need these living conditions, to which they are adapted, today.
3. If there is a lack of an essential factor, this cannot be compensated for by another factor, even if it is also essential. That is to say, if the cactus or fern lacks magnesium, for example, this deficiency cannot be compensated for by potash or calcium.

So someone with a proverbial green thumb does something very simple: they treat the plant according to its evolutionary origins. This is the only way that a cactus or fern in the room, i.e. thrives in an environment that is actually alien to the species. Of course, Sprengel's Law of the Minimum applies to all living beings and any biological system: If something essential is missing, the functions are permanently disrupted. "Species-appropriate" thus stands for the sum of all the essential needs of a particular species. There is no reason to assume that the species concept does not also apply to us humans. On the contrary: The more specifically and advantageously a species—or a subset of it—or a subset of that—has adapted to a particular environment or to certain living conditions, the more important it is to maintain a way of life that meets all these conditions. As a result of evolutionary adaptation processes, Sprengel's Law of the Minimum is thus also crucial for our physical and mental development, as well as for the maintenance of our health and mental power. Thus, any deficiency in an essential factor hinders the development of our frontal lobe and its battery, with all the consequences for the individual discussed previously.

Interestingly, I first learned this elementary "natural law" for acquiring and maintaining the best physical and mental fitness not in medical school, but only in preparation for my participation in the Race Across America. That is astonishing, because this law of nature explains the emergence of most diseases of civilization, which account for the majority of medical activity. If one regards the RAAM as an "experimental life in fast motion" due to its intensity, almost all failed participations there can also be traced back to at least partial ignorance of Sprengel's law of the minimum.[1]

However, deficiencies are not the only cause of frontal lobe developmental disorders or the shrinking of the frontal lobe battery. In a world of abundance that is shaping our cultural development, in particular in the last one to two generations, the brain development of many people in the countries of the global North also suffers from too much. Accordingly, Sprengel's minimum law must be supplemented by a maximum law. The Law of the Maximum, like the Law of the Minimum, is species-specific and thus dependent on the respective evolutionary history. In contrast to the cactus, the fern does not tolerate bright sunlight, whereas the former dies in the long run due to too much moisture. So there is obviously also a maximum for the amount of toxins that a living being can consume. Once this has been achieved, short-term or long-term negative effects can be expected, depending on the poison and species. That sounds trivial. Less trivial is the fact that this applies even to foods and their components that are actually considered healthy or even vital (i.e., subject to the law of the minimum). As a Swiss physician known as Paracelsus already knew some five hundred years ago, "All things are poison, and nothing is without poison; only the dosage alone makes it so a thing is not a poison." Because our "modern" lifestyle almost constantly takes place outside a healthy corridor between minimum and maximum, serious physical and mental developmental disorders or irreparable damage inevitably occur. This "cultural" pandemic of extremes (below the minimum and above the maximum) is orders of magnitude more dangerous than any viral pandemic (including COVID-19) because it is responsible for the majority of all diseases and deaths in our society (yes, even for most deaths from the coronavirus[2]). Nevertheless, both the law of the minimum and the law of the maximum are largely ignored in conventional medicine. When diseases occur, the disease-causing behaviors are hardly ever investigated and, if necessary, corrected instead; usually only the symptoms are treated. To make matters worse, in modern medicine only that which lies significantly (usually several standard deviations) outside this already a priori unnatural normality is defined as sick. Since an unnatural and therefore unhealthy lifestyle is part of normality, the "normal" consequences are often not recognized as illnesses. Even the shrinking of the frontal lobe battery, considered normal in the general population, is ultimately an ignorance of a chronic and completely unnatural neurodegeneration.

Against genetic determinism. It is generally assumed that our mental and physical characteristics are largely determined by our genes. Some scientists even claim, according to a cover story in the weekly newspaper *Die Zeit*, that "we can predict more than ever before from our genetic makeup, even in newborns—intelligence, weight, health."[3] In fact, genes are important, even vital—after all, we would not exist without them—but they are not fate, and instead they endow us with a potential that needs to be used. Our genetic makeup provides a blueprint. A possible picture of the mental capacities of an adolescent child would perhaps be that of a jug. It is wrongly assumed that shape and size are predetermined by nature and that, as parents, we only have the option of providing the best filling. In fact, however, we can also contribute to how big the jug will ultimately be and what shape it will have. We have the possibility to fully exploit the potential given by nature and thus have much more influence and responsibility than we are aware of.

The Emergence of the Human Species

The central question arises as to what living conditions the human genetic program has adapted to and what indispensable minima and not-to-be-exceeded maxima were created as a result. In answering this question, opponents of a species concept for humans often emphasize that evolution would take place constantly, so it is not only a phenomenon of the past, but continues to shape all life on our planet, including ours. In reality, standstill is only an illusion. If living conditions change, random genetic changes in the genetic material of the reproductive generation can ensure that their offspring cope better with the new conditions. Whether a new species emerges from this, however, depends on whether the new variant can reproduce better in the long term and prevail against the earlier one. However, it is unlikely that our genetic makeup will adapt to a lifestyle that has become unnatural for us due to the rapid cultural and technological changes because the prerequisite for these new, "more advantageous" mutations to prevail is selection, which only takes place under certain conditions. The best example of actually occurring evolution is provided by pathogens. Through the widespread use of antibiotics in factory farming (mainly to accelerate the muscle growth of the animals), we are selecting more and more bacteria that have not only developed resistance to a single antibiotic but have multi-resistances to all kinds of antibiotics.[4] However, the antibiotics themselves are not mutagenic—they do not cause changes in the genetic material that cause resistance—and thus do not create new bacterial strains, but are only the selection mechanism that kills all non-resistant germs. If there is no killing of the non-mutated germs, the one new bacterium that happens to have a suitable mutation would have no chance of prevailing against all the others. A genetic adaptation, such as resistance to antibiotics, therefore requires mutation and selection.

In humans, too, an increase in certain resistances, such as against malaria or Lassa fever, can be demonstrated through changes in the genetic makeup, but only because in the disease areas, many people who do not possess a protective mutation die from the disease or are so weakened that they have no or fewer offspring.[5] As a result, the percentage of the protective mutation increases. However, these examples of human evolution are only local in nature and concern only a selective trait such as increased genetic resistance to a likewise mostly local pathogen. However, this does not result in an actually new species with a different metabolism or completely changed behavior. And even if there were such a dramatic genetic change, a global spread is not to be expected, because a selection would also have to take place for that. The new species would have to prevail against "ordinary" humans. Infections, even when they become pandemics, hardly ever lead to heritable resistance, because, as for example with the coronavirus, usually only those members of the world's population who have a very weak immune system die. Most of the others survive. There will therefore be no "enrichment" of random carriers of genetic mutations that would have a natural resistance to such a viral infection. Nor will a new human species prevail that remains healthy despite an unnatural lifestyle, because the "ordinary" human being has usually already passed on his genetic material that is not resistant to this unhealthy way of life. This happens before the diseases of civilization caused by that unhealthy way of life occur. A selection of genetically "new humans" who, for example, cannot be harmed by junk food, therefore, does not take place.

Another reason why the human genome hardly changes or has to change is the fact that humans react to changing living conditions and also an alien way of life with technological or cultural means. Thus, even with severe illnesses often brought on by unhealthy lifestyles, most people today still enjoy relatively long lifespans. The need for a healthy lifestyle is pushed into the background because you have medication and good acute medicine.

Our genetic program will not adapt to global warming either, coupled with a rise in sea levels and the expected flooding of large habitats. We will not grow webbed feet and gills; rather we will build dams or flee inland. Another example: As a result of global warming, plankton in the world's oceans has declined by about 40 percent over the past century.[6] This is dramatic—after all, it produces about half of the oxygen we need to live.[7] But even if, due to our massive interventions in nature, we one day run out of air, a new species of human that might be able to cope with less oxygen will still not evolve. Instead, it is much more likely that a huge market for artificial oxygen will emerge, which we will then supply ourselves with from gas cylinders via breathing masks.

From these considerations it can be deduced that our species could only have evolved *before* it was able to react technologically to changing living conditions. However, there also must have been special conditions for our ancestors to develop the complex mental abilities that characterize us today. Moreover, these living conditions must have been stable over a longer period of time, because the development of the species *Homo sapiens* required many genetic changes that can only have arisen successively, selected over many generations. That means that all those who did not genetically adapt to these new living conditions must have had significantly worse reproduction rates and thus must have fallen victim to selection pressure. In fact, paleontologists found convincing evidence of the near-total extinction of humanity. The cause of this was an ice age that began about 200,000 years ago and lasted approximately 70,000 years, which dried up all the centrally located regions of Africa. This created a kind of bottleneck.[8] Only a few clans of hunter-gatherers who made it to the southeastern coast of Africa made it through and survived the catastrophe. There, they were forced to become fishermen and gatherers and, according to comprehensive genetic analyses, the ancestors of all ethnicities living today. But the narrow coastal strip was not only a refuge, but the scientifically proven Garden of Eden, which above all drove the mental development and genetic selection toward *Homo sapiens*.[9] In particular, the kilometer-long, practically inexhaustible mussel beds became the tree of knowledge, to stay with the biblical metaphor: Not forbidden apples, but seafood, with its extraordinary cocktail of nutrients for human brain growth, paved the way to knowledge and provided the ideal conditions for an explosive growth of the human brain.[10] Countless fossil finds now irrefutably support the thesis that human mental evolution, in particular, benefited from the marine food chain.[11] For example, the highest concentration of the omega-3 fatty acid DHA (docosahexaenoic acid) in the entire brain is found in the frontal lobe. As Figure 11 shows, that proportion increases significantly, especially in the last trimester of pregnancy and the first four years of life. This results in not only an absolute, but also a relative accumulation of this brain building block compared to the brain building block arachidonic acid (AA, an omega-6 fatty acid).[12]

In a certain way, these changes in the ratio of AA to DHA in frontal lobe growth reflect the changed lifestyle of our ancestors, namely the change from hunter-gatherer, who had an AA-rich diet, to fisher-gatherer, where a DHA-rich diet predominated, which promoted their frontal lobe development. But it is not only the phylogenetic change from an AA to a DHA-rich diet that is evident in child development. Quite generally, the development of our frontal lobe was the last and decisive step in our tribal history (phylogenesis), and this is also repeated in our individual development (onto-genesis). Thus, in the context of early childhood brain development, the frontal lobe is the last brain region to experience the decisive growth spurt, which is correlated with a high intake of DHA.[13] This is why the prefrontal cortex, in particular, is the region of the developing neocortex that develops last—and suffers particularly from deficiencies such as DHA during this time window.

It is not only the function and growth of the frontal lobe that are dependent on a sufficient supply of aquatic omega-3 fatty acids, but also the function and growth of its "battery." As Figure 12 shows, the hippocampus volume is directly related to the so-called omega-3 index (the percentage of aquatic omega-3 fatty acids in the cell membrane of red blood cells).

Since all cell membranes—as well as our fat deposits—consist largely of fatty acids and thus form a large storage, the omega-3 index usually changes very slowly, over weeks or months, when one changes their diet.

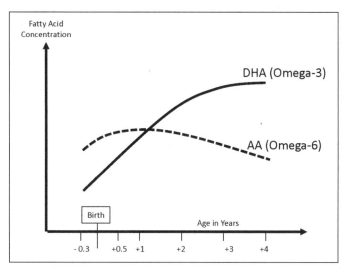

Figure 11: The concentration of the omega-6 fatty acid AA (arachidonic acid) and the omega-3 fatty acid DHA (docosahexaenoic acid) in the developing frontal lobe.

The omega-3 index tells us whether we have consumed too little or enough aquatic omega-3 fatty acids over a longer period of time. Thus, the examination of a drop of blood provides a status report on the supply of these essential fatty acids. Not only because of its importance for the frontal lobe and its battery, but for all bodily functions, such a fatty acid analysis is of enormous clinical significance, which is why I advocate that medical practices perform it routinely. Many lives could be saved as a result. For example, increasing the global average value of the omega-3 index from currently just 4 to 5 percent (marked with a circle in Figure 12) to a value of 8 percent would reduce the risk of sudden cardiac death by about tenfold.[14] Ideally, the individual omega-3 index would even be as high as 8 percent a value from 11 to 12 (also marked with a circle). This would also correspond to that of a fisher and gatherer, who consumed several hundred grams of fish and seafood as his basic food.

The high rate of cardiovascular disease worldwide thus indirectly provides an indication that our genetic makeup is still adapted to a prehistoric diet that was rich in fish and seafood.[15] Further evidence comes from studies of mental fitness in relation to the intake of aquatic omega-3 fatty acids. The more fatty acids the mother consumes from fish and seafood during the crucial growth phases of the child's brain, the better the mental development of a child will be.[16] Even just taking half a tablespoon of DHA-rich cod liver oil daily during pregnancy and breastfeeding showed that, four years after birth, the offspring's mental fitness was on average around 4 percent higher—compared to the children of mothers who consumed corn oil instead.[17] Also the head circumference and birth weight of children of first-time mothers those who consumed DHA regularly during pregnancy are higher than those who consumed little or not at all,[18] although a larger head circumference at birth tends to be associated with better intelligence development in the child. And these effects are long-term, as is the previously discussed protection against schizophrenia. For example, a direct relationship between the DHA concentration in the umbilical cord blood—as a measure of the prenatal supply of this essential brain building material to the unborn child—and the mental performance and memory capacity could be demonstrated in Inuit children, even eleven years after birth.[19] The supply of aquatic omega-3 fatty acids is therefore subject to a clear minimum that should not be undercut if the frontal lobe and its battery are to develop their full potential. According to the results of mathematical calculations from series of such study results and investigations, the following relationship emerges:[20] For every gram of DHA that the mother consumes daily in the child's important development phase, the child's IQ increases by around 0.8 to 1.8 points. Conversely, a loss of intelligence can still be detected at the age of seven if the mother was deficient in DHA before, during, or after delivery.[21]

The lower the absolute and relative proportion of aquatic fatty acids in an adult's diet, the smaller his brain is and the faster it breaks down.[22] According to the law of the minimum, a poor supply of aquatic omega-3 fatty acids has an impact on the entire life. Due to the frontal lobe battery's ability to form new brain cells throughout life, however, it is never too late to ensure a balanced diet. Even in old age, aquatic omega-3 fatty acids support hippocampal growth.[23]

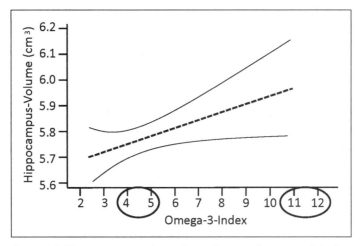

Figure 12: Hippocampus volume depending on the omega-3 index. The dotted line stands for the average value; the thin solid lines show the range of measured values. An index of 4 to 5 is the global average, but 11 to 12 would be natural and optimal for the brain.

Selection of Mental Superiority

Over thousands of generations, the genetic makeup of the initially severely decimated "world population" is becoming more and more adapted to the prevailing living conditions on the southeastern coasts of Africa, whereby a diet rich in aquatic omega-3 fatty acids, vitamins, and essential minerals is an elementary prerequisite for the growth of the brains of our ancestors, but not its cause. A kind of "social selection" could have been responsible for this, because sedentary lifestyle and increasing group size of the fishing and gathering clans required a different way of thinking.

Therefore, there must have been other special circumstances prevailing among our Stone Age ancestors that gave the mentally fitter members of each generation more offspring, so that genetic changes were preferred or selected that provided for a comparatively higher intellectual potential. A kind of "social selection" could have been responsible for this, because sedentary lifestyle and increasing group size of the fishing and gathering clans required a different way of thinking. The former hunter-gatherer communities had to follow their prey nomadically in order to ensure enough food for everyone. In contrast, for the fishing and gathering community, the marine food supply on the coast was practically unlimited, which meant they were able to settle down. The abundance led to a rapid growth of the local population. This trend toward higher population density can be observed among all wild peoples as soon as a clan fed primarily on inexhaustible aquatic sources instead of limited terrestrial sources.[24] The African coastal strip was indeed very rich in food but unfortunately very limited in terms of space. A stable diet with the maritime "brain food" was thus only guaranteed for those clans that had permanent access to the coast. This inevitably led, and perhaps for the first time in human history, to a sense of ownership, an idea that is still largely alien to members of living foraging communities even today.

The larger a group living closely together in a place becomes, the more inevitable are social conflicts that need to be resolved—this is the case today and began to become a new challenge even then. In addition, new land had to be conquered again and again as soon as the coastal home became too small due to an increase in the family. At the same time, the marine food source had to be defended against the intrusion of foreign groups. Anyone who was unsuccessful in these tasks ran the risk of their genetic heritage possibly being completely wiped out. It can be assumed that those families who, due to genetic coincidences, possessed the potential for comparatively higher emotional and social intelligence had better chances of securing their continued existence in the long term. These forms of intelligence made life within the clan easier and ensured that such clans had better cohesion and thus greater power against their enemies. Thus, social intelligence became a decisive genetic selection advantage. This circumstance was certainly decisive for the evolution of cooperative behavior within the respective clan and thus an essential prerequisite for the subsequent "social conquest of the earth"[25] by humans.

Planet Earth's conquest began in Africa and followed waterways.[26] The primary expansion routes were along the Nile's banks, along the waters of the Great African Rift Valley, which stretches for some 6,000 kilometers from present-day Mozambique in southern Africa to Syria in the north, or directly across the sea—and this is surprising.[27] In fact, according to the latest findings, our direct ancestors were already capable of building boats and navigating the open sea 130,000 years ago. The speed of spread was most likely due to the intergenerational growth of the most successful clans, which at the same time ensured that their offspring could always feed themselves with a sufficiently aquatic diet.

People back then were anything but primitive. In addition to the ability to build boats and navigate the sea, *Homo sapiens* developed quite complicated procedures more than 164,000 years ago, such as the targeted heat treatment of stone materials to produce razor-sharp blades.[28] This required experience and planned action as well as a complex language to pass on the acquired knowledge and skills to the next generation. Such technological and cultural achievements require a forward-looking, planning spirit, but also time that can be used creatively, and our ancestors also had that. Because of the rich local food supply, a few hours a day would most likely be enough to meet the demand for seafood and nutritious root vegetables. Afterward there was a lot of time that could be used to relax but also to be creative. This is in contrast to the New Stone Age farmer and cattle breeder,[29] who had to invest much more time just to feed himself. This time, which the fishermen and gatherers of that era had, seems to be missing from us today as well. We often feel stressed, and this is frequently due to a lack of time. Recent findings show what each of us feels: We are more creative and much more effective when we take time to relax.[30]

All these circumstances—optimal brain nutrition, social selection pressure that promoted cooperative action, as well as time to be creative—ensured that over the first fifty thousand years of settlement of the southeastern coast of Africa, which was caused by the Ice Age, the volume of the human skull reached its all-time record.[31] In particular, the prefrontal cortex increased in size compared to other brain regions.[32] We owe our dominance as a species to development: No other species of *Homo* coexisting at that time could have beaten the Sapiens in an Other species vs. *Sapiens* "Frontal Lobe Face-Off." Even our last living relative, the much physically stronger *Homo neanderthalensis*, had nothing to oppose the superior planning, cultural, and social skills of our ancestors. In fact, the different development of the prefrontal cortex is also the essential distinguishing feature between our brain, that is, that of a former fisherman and gatherer, and that of the Neanderthal, whose receding forehead was characteristic.

As a nomadic hunter-gatherer, the Neanderthal did not feed to the same extent as our direct ancestors on seafood, which is so extraordinarily beneficial for frontal lobe growth, but mainly on the meat of ruminants.[33] Thus, the high (thinker's) forehead of *Homo sapiens* became the "trademark" of the only surviving human species. However, this should give us pause, because after all, the majority of people today eat even worse than the Neanderthal once did. While the Neanderthal fed on game and wild plants rich in vital substances, a large part of the world's population eats the unhealthy products of industrialized meat and agriculture. In a way, we have become domesticated hunter-gatherers without hunting or gathering ourselves.

Figure 13: Battle of the frontal lobes. Our closest genetic relative, the Neanderthal, had a significantly less pronounced frontal lobe, clearly recognizable by the receding forehead. In the background is the skull of a modern man.

Species-Appropriate: Formula for a
Robust Frontal Lobe Battery

The human frontal lobe forms new brain cells until the end of adolescence, that is, until about the age of twenty-five. However, this is far from the end of the maturation. According to the results of one study, the volume of white matter in the frontal lobes continues to increase until at least the age of forty-four, and in the temporal lobes even until the age of forty-seven.[34] White matter contains the nerve fibers, and an increase in its volume indicates that even an adult's brain is in a constant state of change, with the complexity of the circuitry continuously increasing. In addition, the frontal lobe battery even has the ability to form new nerve cells throughout life.

Countless scientific studies now show that the living conditions of the fisher-gatherers of that time would still be ideal today for the frontal lobe battery and provide the greatest growth impulses. However, this does not mean that we have to live a prehistoric life to ensure optimal development of the frontal lobe or lifelong growth of its battery. It is entirely sufficient if we integrate the decisive elements in the various areas of life that drove the development of human mental power at that time into our modern lives—always according to the laws of minimum and maximum.

An example: In order for fishermen and gatherers to have enough food, they had to be physically active for a few hours every day. In the process, they always experienced something new that they had to remember. Only those who remembered where they had found something to eat, or where in nature danger lurked, survived. The hippocampus is the only area of our brain that can remember such unique experiences. Accordingly, nature uses hormones that are released directly or indirectly through muscle work and whose primary task is to make us physically fitter as signals that also stimulate hippocampal growth (more on this in more detail later). These growth impulses already ensured that our ancestors had a wealth of experience with the hippocampus and thus the probability that they would survive in the wild. Even today, exercise is an effective means of releasing these hormones and thus expanding the frontal lobe battery capacity throughout life. That's why physically active people are usually more creative and mentally fit than people who are not active. They are also more resistant to stress and suffer less frequently from depression. Overall, exercise protects us from Alzheimer's and many other diseases and thus prolongs our lives.

Almost everything that fishermen and gatherers brought home from their forays as food protected their brains from premature aging for the rest of their lives and promoted the mental development of their children. Thus, as we have seen in the example of aquatic omega-3 fatty acids, nutrition is also of crucial importance for the development and maintenance of mental fitness. However, due to the high level of toxin pollution in the steadily diminishing fish stocks, we now have to use other sources for all food components that are beneficial for brain metabolism. In fact, today you can even live completely vegan, eating neither fish nor seafood, without suffering a deficiency, if you take appropriate alternatives for the essential micronutrients from these food sources. Thus, even without fish, a species-appropriate diet is possible, as we will see.

Unfortunately, many deficiencies in the modern diet favor an inhibition of mental development and damage mental health in the long term. In addition, there are certain food components that also cause chronic frontal lobe weakness due to an excess of food, as well as many harmful substances and toxins, which we will also discuss later. This cultural development of *Homo sapiens* toward frontal lobe weakening probably began around ten to twelve thousand years ago when our ancestors became cattle breeders and farmers (agrarians), because until then all European descendants of the South African fisher-gatherers fed mainly aquatically, with mussels and fish.[35] The no longer entirely species-appropriate diet of the agrarian of that time had consequences. Thus, its brain volume shrank by about 11 percent compared to the fishermen and gatherers of the time, as skull measurements prove.[36]

However, an alien diet is probably not the only reason for this development. At least nowadays, in the social sphere, the release of important hormonal growth factors for brain development and the lifelong capacity expansion of the frontal lobe battery occurs less and less frequently. The reason for this is not least an economic policy system that increasingly individualizes us humans and thus isolates us, but also robs us of the meaning of life. As we will examine in more detail, ironically, modern social media, although they connect us with more and more people digitally, lead to increasing loneliness and social stress. Even toddlers and infants are now suffering from the fatal consequences. Due to perceived and, of course, actual loneliness, there is a chronic release of stress hormones, which not only inhibits the growth of the frontal lobe battery but, as already discussed, also drives its loss of capacity.

In addition, our brain must reflect on all the experiences it has in the active phases of our lives and store them in the long term. This requires restful sleep. Deficiencies are also increasingly emerging in this essential area of our lives. Thus, more and more people are now suffering from poor sleep, with dramatic consequences, especially for the function of the frontal lobe battery. The cause is again chronic stress, but also a lack of active regeneration, either because there is no time for it or even more often because it is no longer used in a species-appropriate way. This is fatal, because time for the muse, which the prehistoric fisherman and gatherer still had, is also important in order to find a new meaning in life again and again throughout the different phases. Only if a person has a deeper reason that makes them get up in the morning will their brain remain functional in the long term—as if the brain senses that it is needed.

In order to fully develop our mental abilities and expand them throughout life, a healthy diet, social and physical activity, restful sleep, and a purpose in life are essential. However, according to the law of the minimum, none of these five elements is more important than the other, or could even replace one of the others. However, each individual element influences the quality of all the others, which is why I have shown them overlapping in Figure 14.

Most of the time, the things we do are partial aspects of various elements of this formula for a high-capacity frontal lobe battery. For example, if we plant an apple tree, it is a meaningful task, combined with physical activity and the chance of a healthy food. Perhaps we do this together with others or share the fruits later, and the action has a deeper social aspect.

Since all these activities or elements of life are of elementary importance, I have embedded them in the larger framework of *time*. This is to make us aware that we have to take time for all these things—and in a healthy relationship. For example, the Race Across America mentioned at the beginning is considered the toughest endurance race in the world precisely because almost all participants choose an extremely unhealthy time ratio between the formula elements of *sleep* and *physical activity* as a race strategy. They almost completely renounce the one, supposedly in favor of the other. Extreme lack of sleep (especially during high physical exertion) is not only unhealthy, it also inhibits regeneration, which massively limits physical performance as the race goes on and ultimately leads to complete collapse for many participants.

Interestingly, the element of *time* is also subject to the laws of minimum and maximum. For example, a lack of time when completing a task leads to stress. The resulting release of stress hormone prevents the growth of the hippocampus, which reduces stress resilience and can lead to a vicious cycle known as burnout. Conversely, you can also have too much time, which leads to boredom, which can be just as stressful in the long run—namely if you feel that your task-free existence is pointless. It is therefore important to be aware of the element of time and to use it wisely, just as important as taking the time to find an answer to the question: Does what I'm doing now give meaning to my life?

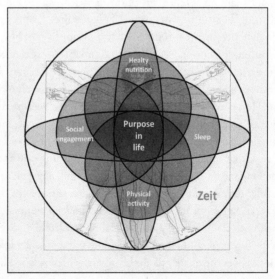

Figure 14: The essential elements for a high-capacity frontal lobe battery.

The GOOD, the BAD and the UGLY

The fisher and gatherer lived in harmony with his nature. It told him how to behave. If our ancestors did something that was contrary to their nature, their chances of survival were reduced. We, on the other hand, can behave completely against our nature for a long time. For example, we can opt for unhealthy food or even drugs. Depending on our profession, we can completely do without being physically active. We can do almost everything that was natural and ultimately without alternative for the fisher and gatherer of that time completely differently without it having immediate serious consequences. Hardly anything kills us immediately. However, constant dripping wears away the stone. Nowadays, most people live such species-inappropriate lives that the chronic loss of frontal lobe battery capacity has become the norm. We are thus living in the age of an increasing, global frontal lobe weakening, but almost no one is aware of this. In industrialized nations, people often think more about the species-appropriate keeping of farm animals than about their own way of life. But even in animal husbandry, one usually limits oneself to the absolute, at least supposed, "minimum of species justice" that the fattened livestock absolutely needs to end up as a cheap sausage on the plate. But why do we behave this way? Individual thinking and behavior are known to be influenced by individual experiences. According to this, the collective thinking and behavior of entire nations is controlled by collective memory, by what we call culture. A possible answer to the central question of why we think we are outside nature and do not behave according to our species—with all the fatal consequences for mental development (and the survival of humanity)—could be provided by a look at our recorded history.

For example, the Judeo-Christian faith shaped the general self-understanding that God created man in his image and that therefore man enjoys a special position similar to God outside of nature. The idea of an almost divine omnipotence of man influences all people who grow up in such a culture, even those who are not believers themselves. With God as our personal creator, according to Harvard psychologist Steven Pinker, it inevitably leads to the "modern denial of human nature,"[37] with profound effects on our self-understanding and the way we shape our lives. One consequence of this is that we elect politicians to power because they suggest omnipotence to us, the boundless belief that man, due to his supposed special position, can create anything.[38] The leaders of the economic and political course convey the conviction that everything that man destroys through his actions—such as his physical or mental health or the global environment—can also be repaired by man as soon as he decides to do so.[39] The impossibility of curing diseases caused by a species-inappropriate lifestyle with medication, or the daily bad news about environmental destruction and species extinction, show that this is not true.

Furthermore, there is the erroneous belief that man—like a god—can control nature and shape it according to his will. Purely economic policy interests are always in the foreground, more or less blatantly, often justified by the argument that the world cannot be saved if one does not generate enough money to do so oneself. This sentence could also come from the doctrine of faith reshaped by Western culture. Rolf Schieder, professor of practical theology at Berlin's Humboldt University, is convinced that "The Bible does not demand poverty from everyone." In other words, there may also be rich people, in fact there must be. Because, Schieder continues, "how can you care for the poor if you have nothing yourself?"[40] The belief in the omnipotence of man is thus manifested in the omnipotence of money.

The international title of the third and final film in the "Dollar Trilogy" by Italian director Sergio Leone (1929–1989) fits in with this. In my opinion, *The Good, the Bad and the Ugly* describes quite well the situation in which humanity now finds itself. Very few people know what would be good for us (the GOOD), or they ignore it. As a result, the bad prevails (the BAD). Economic interests and economic policy decisions and measures ensure that this remains the case. That's the ugly thing about it (the UGLY). As a result of this culturally based and widely accepted economic policy course, all people are the ones who suffer, especially our children and grandchildren. Their future is at stake. As a rule, the goal of state legislation is not its species-appropriate development, but an economic development. Educational measures focus more and more on later professional life instead of the natural development of physical and mental potential. The growth of the gross national product seems to be more important than the growth of the frontal lobe battery, with all the catastrophic consequences for physical and mental health that are now known to you. For example, political parties, some of which, interestingly, describe themselves as "Christian," repeatedly pass laws that contribute to further destroying our livelihoods and jeopardizing the health and future of our children.[41] At the same time, they delay or even refuse to pass laws that would protect us and our children from health hazards. Many serious and at the same time frightening examples will follow. I will introduce it with the term "The UGLY." But first, we'll look at what would be natural and why it's worth investing in all the elements of healthy living—for yourself and for your children.

CHAPTER 8

ON THE MEANING OF LIFE

The mystery of human existence lies not in just staying alive,
but in finding something to live for.
—Fyodor Dostoevsky (1821–1881)

The Search for Meaning

In the stormy and rebellious phase of my youth, I questioned everything, even my own existence. Am I real or am I just an illusion of myself? And if I really exist, what is the meaning of my existence? An essay that particularly moved me at the time was Albert Camus's The Myth of Sisyphus.[1] *It begins with the words: "There is but one truly serious philosophical problem, and that is suicide. Judging whether life is or is not worth living amounts to answering the fundamental question of philosophy." Now, I wasn't suicidal, but the search for answers to such an elementary question was exactly my thing, and even today I start from extreme situations when questioning facts in order to get to the core of the problem more quickly. According to Camus, human existence is both hopeless and absurd. God is dead and life as a whole is meaningless, he is convinced. Among others, the German philosopher Friedrich Nietzsche (1844–1900) probably brought him to this conclusion. We would be looking for meaning in a meaningless universe. That was the basis of his thinking, and— philosophically speaking—you can't sink any lower. But this fundamental thought, according to Camus, also sets us free. Free from all supposedly divine, cultural, or social missions. It is only we ourselves who can answer the question of the meaning of existence. Because if there is no absolute, overarching meaning, then at least the option of a relative, personal one. We just need courage and self-confidence to seek it, that was his message. Consequently, we should "imagine Sisyphus as a happy man." The Sisyphean task of repeatedly rolling a heavy stone up a mountain, although in absolute terms a meaningless endeavor, without purpose and without prospect of success, is nevertheless a fulfilling task and thus the personal meaning of Sisyphus's existence. However, I saw a logic problem in this, because Sisyphus had not chosen his "life's work" himself. There is no trace of self-determination in the search for meaning. In addition, I could not find anything positive in the circumstance of the endlessness of divine punishment. This raised the question for me whether life has to be limited in order to have meaning for the individual.*

Searching for the meaning of life is a very special characteristic of being human. Philosophers and representatives of all religions have been occupied since time immemorial with finding a satisfactory answer to this for all people, but none has been convincing so far.[2] Whether life itself or whether evolution has a purpose, no one can say. There are theories, but nothing more. More important, however, seems to me to be the more specific question of whether *our* lives have meaning for *us* as individuals. This can actually be answered. A simple question can help: Do we have an important reason to get up in the morning? Having to use the bathroom or simply having a good night's sleep and an appetite does not count here. Rather, it is about whether a task that we consider meaningful is waiting for us, whether we see meaning in what we are going to do. If that is the case, then there is something that in*spire*s us, that enlivens our *spirit*, and we have found our personal meaning in life, at least in this phase of our lives.

The thrilling experience of meaningful tasks is essential for the networking of the nerve cells in the hippocampus that are newly created every day. After all, the life-sustaining task of these young nerve cells is to store new emotional experiences. If we or these new nerve cells do not experience anything new that evokes an emotion in us and therefore needs to be noticed, there is no reason for their continued existence. As a result, the unneeded nerve cells commit preprogrammed suicide (apoptosis) and can no longer contribute to strengthening the frontal lobe battery. However, if we have a personal purpose in life that we pursue, then we also experience something worth remembering for the hippocampus, and the capacity of the battery increases every day. Meaning in life is therefore the central element of the formula for a strong frontal lobe battery outlined in the previous chapter. All other elements of the formula are also needed, but they are always subordinate to the individual meaning of life and are usually essential for its implementation.

Meaning in life is therefore vital. Without meaning in life, we feel lost. In this respect, one could actually imagine Camus's Sisyphus as a happy person. But in the real world, the meaning of life is subject to constant change, because we and the world around us are changing (unlike Sisyphus). But we also have to change in order to always be able to break new ground. But in which direction? Our personal answer to the question of meaning helps us to reorient ourselves, like a compass, again and again. Those who seek meaning live happier—at least the next day, as the results of a psychological study show.[3] One explanation for this could be that those who search for meaning are more likely to perceive meaningful information from the environment.

The enthusiasm when you have new experiences usually also causes positive stress, also known as eustress. Both a moderate dose of stress hormone, as is released during eustress, and new emotional experiences are basic prerequisites and thus decisive for the new hippocampal nerve cells to connect synaptically, thereby surviving and allowing the frontal lobe battery to grow.[4]

If hippocampal growth is regarded as an indirect measure of a meaningful life, Camus must have been mistaken in his interpretation of Sisyphus as a happy person. Sisyphus may have a divine task, but he himself must recognize it as meaningless (this is exactly where the punishment lies). It must also be stressful for him, since it is unsolvable; moreover, due to its repetitive character, it does not provide any new experiential content. Even if he were to form new hippocampal nerve cells under this task (after all, he moves a lot), they will never have a new experience to remember due to the monotony of his actions. They are not needed and will therefore die. I'm pretty sure that if Sisyphus's hippocampus were to be measured (he would still have to be busy rolling stones due to the eternity of his punishment), it would be completely atrophied. As a result, he would be a highly depressed man with a high risk of Alzheimer's disease, without self-confidence, and, like Henry Molaison, he would be without a personal memory of himself.

The Question of Meaning from an Evolutionary Biological Point of View

We are creatures of an evolutionary biological process that began a few billion years ago and has since produced new living beings. This will continue to be the case. What could be the answer to the meaning of life from an evolutionary biological point of view? There may be no meaning, life simply exists, it does not pursue or serve a higher goal. The answer to the question of the personal meaning of one's own life is different. From the point of view of nature, the purpose of every living being is to pass on its genetic material. For example, all body parts and functions serve to at least secure one's own life and ultimately to reproduce. They are therefore meaningful. However, it can be assumed that evolution is completely indifferent to whether you yourself see it that way and starting a family is your wish, or whether you pursue completely different life plans.

Evolution is only one mechanism. We humans, on the other hand, are capable of observing ourselves and thinking about ourselves and our future. Our view of our life planning does not necessarily have to coincide with the evolutionary imperative of "be fruitful and multiply." We can pursue completely different life plans. Nevertheless, the basis for a meaningful life for most people is to have children of their own. One reason for this is the fact that our genetic program, as with all other living beings, is geared toward generative success. Hormonal agents that influence our body and also our brain ensure that we have this evolutionary need to feel "priority," to feel a meaning in life and ultimately to pursue it. When falling in love, as well as during lifelong emotional attachment to one's own child, hormones are constantly released that stimulate growth of our frontal lobe battery and at the same time ensure that we feel good about it. If this were not the case, we would have died out long ago.

Since humans, in contrast to many other species, do not rely on quantity for reproductive success (meaning we do not have many offspring, which in many other species are even left to their own devices after birth), but on quality (meaning few offspring who are intensively cared for over several decades), the biological success strategy of our ancestors was already based on the collection and use of experiential knowledge and ultimately also on its transmission within close, cooperative communities. Those who mastered this well had a higher chance of success in finding a partner and successfully raising their children together: passing on their genetic makeup, preserving it, and thus also spreading it. This strategy, in which one asserts oneself with social and emotional intelligence and takes care of the family offspring until at least the second generation, has been very successful by evolutionary standards—about eight billion people are the current result of this evolutionary process. From this, our genetic program was selected over countless (successful) generations, which includes mental growth and the ability to develop mentally throughout our lives.[5]

Scientists refer to *generativity* as the ability to be aware of the mutual dependence of generations on one another, to derive responsibility from this, and to be caring toward people of another generation. Generativity has enormous potential for creating meaning for individuals, but also for society. The simplest form of generativity is to have children yourself and raise them. However, since our close relatives also possess a large part of our genes, especially the siblings, "indirect reproduction" is also possible, for example by taking care of the children of the extended family as an uncle or aunt and thus ensuring the reproductive success of the common genetic material. In the end, if you follow this genetic logic, all people are related to each other. Even though humanity seems very diverse, on the evolutionary timescale, it only recently went through the bottleneck of its almost complete annihilation (see previous chapter), so the genetic differences between different ethnicities are insignificant. Accordingly, if one applies the evolutionary imperative to the preservation of the human genome itself, it becomes understandable why people can see their purpose in life in helping others who do not directly belong to their own family through social activities.

In the search for personal meaning, there are correspondingly many options. "There is no objective meaning in anywhere; it is we ourselves who ascribe meaning to situations," explains Tatjana Schnell.[6] According to the Innsbruck psychologist and meaning researcher, most meaningful activities convey the feeling that they continue to have an effect on one's own life in future generations. You can also adopt a child to pass on your experience and knowledge, be artistically active, or even write a book—everything contributes to the feeling that something of you lives on when you are no longer there.

Let (Not Only) the Children Play . . . (Santana)

Evolution may therefore have no meaning and as such may not pursue a deeper purpose, but every living being strives for the evolutionary purpose of life to maintain itself (life drive) and to be able to reproduce (sex drive). In this way, all genetic programs were optimized in terms of evolutionary history. If humans are possibly the only species on this planet that are able to not accept this evolutionary imperative for themselves and to ask themselves the question of meaning again and again in the course of their lives, then they do so also in view of the awareness of the finiteness of their existence. But what about small children for whom this problem does not yet exist? In their first years of life, children do not look for the meaning of their lives, they simply live—and play. But this also serves the evolutionary purpose of life, because play is usually a safe preparation for later adulthood, where one can take responsibility based on the experiences made through play. Interest in the game was therefore always subject to evolutionary selection pressure, which changed the human genome. Playing releases corresponding hormones and neurotransmitters and is therefore fun—and this ensures that you enjoy doing it again and again. Those who played a lot due to the hormonal driving force gained a lot of experience and were therefore more likely to pass on their gene program. Thus, over a long evolutionary period of time, there was a genetic selection process to be as playful as possible, to be "child-friendly." This need has not changed to this day.

It is important for a child's development to gain their own experience and in this way to conquer their own world (view). Children succeed in doing this through play, which is why, in my opinion, this is exactly the purpose of a child's life. Play is considered so important for a child's development that it has been recognized by the United Nations High Commission for Human Rights as a right of every child.[7] Play—and this means free, unstructured time—is also of essential importance for the cognitive, physical, social, and emotional well-being of older children and adolescents.[8] Even as adults, we do not lose the need to play and learn playfully, says Scott G. Eberle, editor of the *American Journal of Play*, even if noncompetitive play (which would be just as good for this) is frowned upon in our society as soon as serious adult existence has begun and one no longer has time to "waste." This is a problem, because according to studies by Stuart Brown, psychiatrist at the National Institute for Play and author of the bestseller *Play*,[9] we need play as urgently as oxygen (even if there is a lack of this, the brain shrinks), and the opportunity for play is all around us. Yet, according to Brown, we "hardly notice our need to play, or we don't appreciate it until we lack it."[10] This is surprising and a pity, because there are many things that Brown says are ultimately play: Art, books, cinema, theater, but also daydreaming and even flirting. For example, this type of play can help couples keep their relationship young and strengthen the emotional bond. So even in adults, it still serves the evolutionary principle of keeping the partnership relationship healthy, thus leading to family success.

But although these findings are based on scientifically proven facts, fewer and fewer adults are playing—there is a lack of time and leisure. Children also get less and less opportunity to develop in free play. The focus is on education in line with the economy instead of species-appropriate development. Even admission to the "right" daycare center could help decide whether the child will one day have a "career." At least that was the concern of an acquaintance of mine who felt this social selection pressure. As soon as the date of birth was fixed, it was therefore necessary to secure a place for one's future Nobel Prize winner or industrial leader in the "elite daycare center" that promised the following: The child should master at least two languages relevant to economic policy before entering school and become a "virtuoso" on the piano or violin. In addition, the daycare center should be renowned for initiating career paths. To make it all work, the parents hover like helicopters over their overscheduled child. There is no time for casual play and free development—social skills suffer. Because without free play, there is no free thinking—but parents also have less to fear that their offspring will one day question this kind of education and rebel. The fierce competition for the coveted places in the elite daycare centers shows that my acquaintance is not an isolated case. Only children with a good reputation have a chance. To function in this system, you have to become part of it quickly, efficiently, and as early as possible. However, the constant stress, combined with a high pressure of expectations, does not cause the eustress that is beneficial for the child's brain development, but usually only harmful distress. That's the irony of it, because mental resilience in a competitive professional environment is the real goal of this ordeal.

But it is also a constant struggle for the parents themselves. Conflicts with the child and also between the parents are usually unavoidable and exacerbate the problem on all sides. According to the American psychologist Marie Hartwell-Walker, unstructured, free play is becoming increasingly important for the development of our children precisely because it is being lost more and more in today's world. According to Michael Patte, an expert in child development, recent research suggests "children should experience twice as much unstructured time as structured play experiences, highlighting the benefits of unstructured play for the child's overall development."[11] Sergio Pellis, a behavioral researcher at the University of Lethbridge in Alberta, Canada, goes one step further. In his opinion, time in the classroom, which is generally considered the most important for the development of the brain, may be less important than time on the playground: "The experience of playing changes the connections of neurons at the front end of their brains, and without play experience, those neurons aren't changed."[12] Especially those nerve cells that develop during early childhood through free play in the prefrontal cortex contribute to optimally networking the brain's executive control center. Only in this way can it develop its crucial function in regulating emotions, creating plans, and solving problems. "Playing," Pellis adds, "prepares a young brain for life, for love, and even for schoolwork." But all of this only happens in free play: without a coach, without a referee, and without predetermined rules. Children have to negotiate for themselves which game they want to play and which rules they want to follow.

In addition, children become aware of their own feelings and those of others in free play and learn to deal with them. To do this, they often act out things they see and experience in their everyday lives. They process their emotions in the game and learn to express them in the process. Only in this way, in free play, does the brain build new circuits in the prefrontal cortex to control these complex social interactions, according to Pellis.[13]

This seems to be a universal principle. Even in the animal world, for example, the game is not about a young dog or a young cat becoming a better hunter. In fact, the game does not use the ability of fighting or killing (predators can do this by nature), as experimental psychologist Jaak Panksepp of Washington State University, who died in 2017, discovered.[14] Panksepp recognized that play has the same purpose for animals as it does for humans—and that is to "build prosocial brains that know how to interact with others in a positive way,"[15] to learn and gain experience in an informal way. In fact, the play behavior in the entire animal kingdom is astonishing. Rats and monkeys, for example, adhere to rules similar to those of human children when they set them themselves. For example, the players take turns, are fair to each other, and do not inflict pain on each other. In addition, as Sergio Pellis discovered, playful fighting in animals also ensures that they develop greater psychological resilience early on.[16] This means that in playful preparation for later life, their hippocampus and at the same time their social and emotional wealth of experience grows.

In humans, the same playful effect leads to better academic performance.[17] Thus, it could be shown that the strongest predictor of academic success in the eighth grade was social skills in the third grade. Since self-confidence, creativity, and memory also become stronger with a growing frontal lobe battery, many positive effects combine in a synergistic way. Also, the ability to put oneself in the place of others (but also oneself) and to recognize their feelings (as one's own) as the origin of their (and one's own) behavior, which is called "Theory of Mind" in cognitive science, is better developed in free play. Presumably, playing is even a prerequisite for this; if there is not enough time to play, there is a correlation to a lack in physical development, emotional maturation, and the flexible thinking, i.e. everything that would actually be necessary to enable the child to live a successful and, above all, happy life. That would then also be the kind of success that cannot be measured with money. Breaks in everyday school life, and thus the supposed non-learning, are at least as important as structured learning. The American Academy of Pediatrics, a community for the well-being of children to which most US pediatricians belong, also considers sufficient break times with free play to be a "fundamental component for normal child growth and development."[18] Therefore, breaks should not be reduced, either as a punitive measure or for academic reasons. The National Association for the Education of Young Children concluded, after evaluating relevant studies, that sufficient breaks and playing outdoors are crucial for school success.[19] For example, it could be shown that students who do not regularly participate in breaks have increased difficulty concentrating on tasks in the classroom, are restless, and are more easily distracted. Breaks give students the opportunity to develop and improve their social skills, just like free play in their free time. During the break, students learn to solve conflicts and problems, to work with others, and to negotiate without the intervention of adults. In addition, breaks serve as a developmentally appropriate strategy for stress reduction. This is of enormous importance because, unfortunately, according to the consensus report of the US Committee on Physical Activity and Physical Education in the School Environment, today's society places "many pupils and students under pressure due to academic demands, family problems and peer pressure under considerable pressure and stress."[20]

Meaninglessness in media. Even in early childhood, a large part of life takes place with a fixed gaze on the screen and negatively influences the parent-child relationship, as a Leipzig research team discovered.[21] They investigated screen time, which includes television, game consoles, and computer and cell phone use in mothers of two- to nine-year-olds. The scientists found that children of mothers who spent more than five hours a day in front of these devices were also more likely to look at screens for very long periods of time, usually over two hours a day. Children take their mother as a role model or simply have no other choice, because the screen is becoming a supposedly free babysitter for more and more parents. But this comes at a price: Long screen time causes symptoms in children, but also in their mothers, that are known from ADHD. Hyperactivity, impaired attention and often poor social behavior indicate insufficient growth of the frontal lobe battery.

Natural Meaning of Life in Old Age

Pablo Picasso once put it this way: "The meaning of life is to find your gift. The purpose of life is to give it away." The most important gift in the time of the prehistoric fisher and gatherer was most likely experiential knowledge. Thus, the younger members of a clan learned from the experienced elders which things were edible and which were poisonous. Also how tools could be made with simple means or how diseases could be treated with natural means. These experiences were essential for the survival of the younger ones—but also for the older ones. Eventually, the responsibility that they had through the passing on of their knowledge became a task and a meaning in life until old age. Therefore, from a purely evolutionary point of view, grandparents are the ideal babysitters. Not only the parents and their children benefit from this, but so do Grandma and Grandpa. They are less lonely and isolated, and most importantly, this task boosts their self-esteem. Caring for the second generation allows their frontal lobe battery to grow measurably. Retirees were studied who used their free time to educate children and young people as part of the Experience Corps, founded in the US.[22] Michelle Obama, patron of the initiative, recognized that "so many seniors are at the peak of their lives and we need their skills and talents to pass them on to the next generation." With taking on caring for children or educating children, the volume of the hippocampus of the retirees also grew. Compared to those who preferred to stay at home and showed no interest in getting involved as "assistant teachers," the frontal lobe battery grew by up to 1.6 percent over the two-year observation period.[23] This, instead of shrinking by about 1 percent annually, as has unfortunately become the norm in society as a whole, where (not only) older people too often no longer feel a sense of purpose in life.

Playing is also vital until old age and allows the frontal lobe battery to grow. For example, regular chess or card games reduce the risk of Alzheimer's by a factor of three, as was found in a twenty-year leisure study of pensioners on the subject of Alzheimer's prevention. This is compared to those seniors who rarely used their minds in a playful way.[24] To prevent Alzheimer's, the playful challenge must be complex and touch on many facets of how we think and feel. When, as in chess, a spatial memory is stored, the hippocampus is challenged. In addition, the game requires creativity, evaluating ideas and making decisions. In addition, it usually takes place in a competitive and complex context due to social contact, which arouses emotions. The entire brain is therefore highly active. This is also why playing cards can trump Alzheimer's disease.[25]

Play wins, Alzheimer's loses. So-called Alzheimer's mice develop this specific form of dementia, on the one hand, because they are implanted with mutated human genes in their genome, which also accelerate the disease process in humans, and on the other hand, because they are confined to narrow cages where they have no opportunity to play. However, if the laboratory mice are not kept in typical mental deprivation, but in an environment that is stimulating for them, the outbreak of the disease is completely prevented.[26] The genetically modified mice remain healthy and are even mentally fitter than their siblings who do not carry any of these "Alzheimer's genes" in their genome, but who live a "couch potato" existence in unnatural laboratory cages. To prevent Alzheimer's disease, it is therefore completely sufficient to build the animals an adventure playground and let them play with other animals.[27] Not only is an accumulation of the so-called Alzheimer's toxin (β-amyloid, which is produced in large quantities as a result of the genetic defects and then clumps together) prevented, but hippocampal neurogenesis is also very strongly stimulated. The positive effects of playing were demonstrated even in very old Alzheimer's mice, even when the disease had already broken out or progressed in them.[28] One thing is certain—even if it can be assumed that the mice, in contrast to humans, are not aware of it themselves, their meaning in life, as with us, consists of lifelong playing and learning and not to sit bored in a laboratory cage.

"Finally being able to do what you want" is the motto that many associate with the dream of retirement. But due to the temptation of doing nothing, this dream ends far too often in front of the television and with the development of dementia caused by this idleness, all too often in a nightmare. Every hour that we sit in front of the screen each day increases the risk of developing Alzheimer's disease by about 30 percent.[29] The average German spends three and a half hours doing this, making television an important cultural risk factor for Alzheimer's or for the depletion of the frontal lobe battery.[30] Ultimately, this probably applies to every form of inactive screen time, because it is not television itself, but the fact that we waste valuable hours that we could use for meaningful, mentally stimulating activities—or for sleeping, because we supposedly never have enough time for that either.

TV rots your brain. In particular, executive functions decline the longer and more frequently one watches television, even as a young adult, according to the results of a longitudinal study by the Northern California Institute for Research and Education.[31] But what are the causes? Is it because of the program, which only entertains without making the brain think? Or because it prevents the viewer from mentally stimulating activities, so that in the long run the circuits of the nerve cells in the brain are reduced and their intelligence is diminished as a result? Perhaps it is not only the lack of sleep caused by a late-night program, but also the TV-induced lack of exercise. It is well known that a lack of physical activity is associated with poorer brain performance. This was also a result of the study: The more time, on average, that the twenty-five-year-old test group spent in front of the TV at the beginning of the study, the less time they took for exercise. Television time and lack of exercise actually synergistically drove the functional degradation of the executive brain over the course of the two to three decades that the study ran. Little movement reduced their thinking and processing speed, as did a lot of television. The consequence of this mass cultural phenomenon of meaninglessness is thus a continuous decline in the ability to use System II efficiently.

Minimum and Maximum of Meaningfulness

In 1908, two American psychologists, Robert Yerkes and John D. Dodson, published the Yerkes-Dodson law. They found that our mental (and ultimately also our physical) performance depends on the level of arousal or stress.[32] The stress level, graphically depicted, runs in the form of an inverted U: there is a minimum and a maximum of stress, a lack of stress or distress, under which we cannot think and function as well as we would at an intermediate level, under eustress. Over a hundred years later, Dutch brain researchers Dirk-Jan Saaltink and Erno Vreugdenhil also recognized the same "U-shaped" relationship between stress level and adult hippocampal neurogenesis.[33] This works best under eustress, exactly between the two extremes. I have added to the relationship between arousal (stress levels) in Figure 15 to include its importance in the development of Alzheimer's disease and depression, as well as protection against these diseases in eustress.[34]

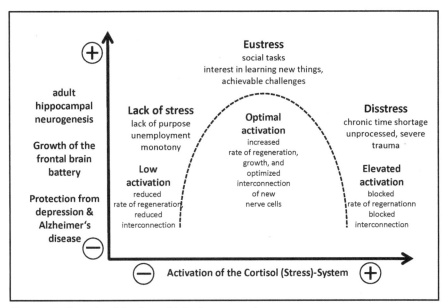

Figure 15: Growth of the frontal lobe battery under stress.

Eustress provides a moderate level of stress hormones, which tells the frontal lobe battery that we are having new experiences. Early in the evolution of the animal kingdom, it was vital to use it to secure the memory of new experiences: Its growth is stimulated without damaging it by too much cortisol (as with distress). Eustress is therefore the quintessence of meaning and positive meaning in life. Even from an early age, eustress, as it occurs naturally in free play, is important for the development of personality, the building of long-term relationships with other people, and ultimately for a creative way of life. Without the stimulating effect of eustress, which we experience in the active, self-determined coping with tasks, newly formed brain cells would not create meaningful and thus stable memories. Eustress thus increases, among other things, psychological resilience, self-esteem, creativity, and memory performance. Last but not least, it protects against depression and Alzheimer's disease.

However, with the realization that the hippocampus is also the frontal lobe battery, this understanding has practical consequences not only for the prevention and therapy of these clinical pictures, but because the development and lifelong growth of the hippocampus also has a fundamental influence on how we perceive ourselves and the world around us, whether we go through life with high or low self-esteem, with which thought system, I or II, we make our decisions every day, and whether we are creative, open to new things, or trapped in our daily routines and stereotypical patterns of thought and behavior. Exactly the opposite happens when you feel that you are no longer needed or when you are actually no longer needed. Complete freedom from stress or the fundamental avoidance of stress is therefore unhealthy and not conducive to maintaining our mental fitness.

However, the law of the minimum applies to the meaning of life, but so does that of the maximum. Those who constantly pursue too many, albeit supposedly meaningful, goals and thus demand too much of themselves often suffer from chronic distress. Continuous demands become excessive demands due to lack of time and are just as harmful to successful neurogenesis as too much time, i.e. an underchallenge. Mental trauma can also cause a chronically high stress level if one does not take the time to process it. This not only permanently blocks hippocampal neurogenesis and thus prevents the frontal lobe battery from growing, but even causes it to shrink. People with post-traumatic stress disorder (PTSD) therefore have a significantly smaller hippocampus than those who have not experienced trauma.[35] However, the risk that a traumatic experience will lead to PTSD is greater the more disturbed hippocampal neurogenesis was previously. After all, it reduces psychological resilience, which promotes the development of PTSD as well as depression. Whether something traumatic happens to us is not under our control, but being prepared for it with a strong frontal lobe battery makes dealing with it easier.

Negative self-reinforcement is found not only in depression, but also in burnout. Distress in the workplace or often an overload in the private sphere (as well as the COVID measures, if they destroy the livelihood) inhibit new nerve cell formation in the hippocampus and thus even lead to a loss of frontal lobe battery capacity in the long run.[36] This reduces psychological resilience, which then often leads to severe depression. Social Occupational Stress, the scientific term for it, should therefore be abbreviated as SOS. Conversely, however, there is also evidence that a smaller hippocampus (an indication of a preexisting neurogenesis disorder and progressive neurodegeneration) shifts occupational or private stress from eustress toward distress by reducing mental resilience.[37] Regardless of whether it is cause or effect, in the end one lives in a vicious circle that increasingly damages the frontal lobe battery: Its capacity is decreasing.

Unfortunately, that's not all. It has been shown that in addition to the hippocampus, the frontal lobe itself, and here in particular the prefrontal cortex, reacts highly sensitively to chronic distress.[38] This leads to neuronal pathway stunting, and thus to reduced connectivity. The result is an increasing mental rigidity. Without a sense of purpose in life, the frontal lobe and its battery lose performance, and one leads—without wanting to and with ever greater difficulty in changing this—a life that seems to have no real meaning. The vicious circle can then only be broken, if at all, with external help.

Sense of Survival

People learn from people. Humans need people to survive and reproduce. Therefore, for the overwhelming majority of people, interpersonal relationships are the authoritative source of a meaningful life.[39] This was tested by scientists using a virtual game. In *Cyberball*, three cartoon characters represent three participants who are connected to each other via the internet. The person assigned to the cartoon figure possessing the ball on the computer screen can decide with a mouse click to which of the other two cartoon figures or people they throw the ball. In the experiment, however, two players were computer-generated, who either threw balls to the only human player, the test subject, or deliberately excluded them.[40] Test subjects who initially received a few balls, but then gradually none, developed, compared to those who were constantly thrown balls and thus in a certain way "liked," a significantly worse attitude and felt less connected to their teammates. In addition, test subjects who were psychologically predisposed sometimes showed severe and long-lasting aversive reactions.

They avoided social contacts after the "antisocial" game, most likely to protect themselves and to not have to experience such exclusion again. Surprisingly, almost all participants who were socially "shunned"—even those without a previous burden—felt that their existence was less meaningful after the rigged game than those who "belonged." This reaction is easy to understand—we find meaning in social relationships because, in evolutionary history, survival was not possible without the protection of or cooperation with other people.

The will to survive, which is thus indirectly reflected in social needs, becomes the direct, all-transcending meaning of life in extreme, life-threatening situations. This was described by the Austrian psychologist Viktor E. Frankl in his book *Man's Search for Meaning*. Despite all the horrors that the Jewish scholar experienced during the Nazi regime, he came to the conclusion, in his own words, "that even in places of greatest inhumanity it is possible to see meaning in life."[41] For Marc Wallert, who came from Göttingen, a life-threatening event also became meaningful. Wallert was a successful, but by his own admission, not too happy management consultant. He kept asking himself about the deeper meaning of what he was doing. But when he was kidnapped on Easter Sunday in 2000 with twenty other diving tourists on the Malaysian island of Pulau Sipadan and then abducted to the Philippines, he found the answer to the question of the meaning of his life in the face of immediate threat: "I suddenly knew exactly what I was fighting for: for my survival and that of my parents. I imagined what I would do when I was a free man again. This vision gave me a lot of strength," he reported after the 140-day martyrdom.[42] It wasn't the jungle that broke him; rather, about five years later, the everyday life that he continued to perceive as "meaning-less" drove him into burnout. After coming to terms with this paradox, Marc Wallert changed his life fundamentally.[43] In the meantime, he passes on the lessons from his self-experience as a resilience trainer and wants to "make a tangible contribution to ensuring that other people are healthy and successful and remain so."[44]

Fortunately, most people are spared such extremely threatening situations as Frankl and Wallert had to experience. Nevertheless, having to find meaning in life in pure survival is not uncommon, for example in the case of life-threatening illnesses. Here, too, a strong will to survive can help one to recover. But even if life is not acutely in danger and forces you to survive alone, to make it a priority, it basically helps to keep facing new challenges in order to stay healthy.

To see meaning in life is to have a purpose. It acts like "an anchor that we cast into the future," says American meaning researcher Michael Steger of Colorado State University; "it keeps the future alive within us."[45] People who experience meaning in life live more future-oriented and therefore usually more consciously than people who are not guided by a deeper meaning—and also healthier: They exercise more, eat better, and use drugs less often.[46] Psychologist Tatjana Schnell calls this a meaning-based orientation of life. The experience of meaning often includes the feeling of having significance oneself, of sensing that one's own actions make a difference, for oneself and for others. This increases self-awareness as well as the openness to experience and learning new things—and positive experiences increase both the cell division rate and the success of adult hippocampal neurogenesis.

Life's Purpose Steers the Genetic Programming

Feeling meaning in life is crucial for the hippocampus to grow and psychological resilience to be strengthened. Having a purpose in life is our most important elixir of life and even makes us live longer—as if our body feels that it needs to stay healthy because our mind senses that it will be needed for longer. The risk of Alzheimer's is also reduced—just as if our brain knew that it was necessary to remain intact. Thus, people with a comparatively high sense of purpose are about two and a half times less likely to develop this most common form of dementia.[47] Since Alzheimer's is causally linked to a shrinking frontal lobe battery and protection against this form of dementia correspondingly in its growth, the question arises as to how meaning in life activates adult hippocampal neurogenesis. Or vice versa: how a lack of meaning in life causes the hippocampus to atrophy.

Feeling a sense of purpose in life has a direct influence on the activity of certain genes in our genome. For instance, in individuals who lead a life oriented toward positive goals, genes that code for inflammation-promoting messengers are less active, resulting in the production of fewer of these agents.[48] This is beneficial because chronically elevated blood levels of these messengers are associated with accelerated physical decline and increased mortality.[49] Additionally, it has been shown that some of these messengers not only inhibit the growth of the frontal lobe battery but also specifically accelerate its neuronal degradation.[50] Blocking their effect or reducing their activity can stimulate the impaired hippocampal neurogenesis again, with all the positive consequences for mental health.[51] In the case of eustress, however, the production and regulation of the release of the stress hormone cortisol to a moderate level, is needed for the frontal lobe battery growth.[52] Meaning in life and the associated eustress thus serve as a stress buffer. However, if there is a lack of meaning in life and one suffers precisely because of this from sickening distress with high cortisol release, all body cells, including those of the brain, age significantly faster.[53] The increased cortisol levels in chronic distress also increase the production and release of the inflammatory substances mentioned above, which exacerbates the problem.

It therefore makes perfect sense to strengthen the sense of meaning in people who lack meaning in life by means of therapeutic intervention. It is about helping to focus on meaningful personal issues through self-reflection, to reflect on experiences in the past and to draw strength from them for new challenges. It is important to create a new perspective and in this way identify meaningful activities for the rest of life.[54] In this way, the crisis of meaning can be overcome and thus the quality of life can be increased in the long term.

"As long as I don't implement the things that I see as meaningful, I won't have a sense of meaning," says Tatjana Schnell.[55]

What is the personal meaning of life? As we have seen, there is no general answer to this question. But you will recognize the meaning of your life when you have found it. And you will find it if you look for it. My own search for meaning always begins with the question that I—fictionally lying on my deathbed—retrospectively ask myself: "Wouldn't I have preferred to spend my valuable time on something else, or was it right to have done this or that?" Although this way of finding meaning may be a bit macabre, the effect is immediately noticeable.

CHAPTER 9

THE SOCIAL PERSON

There is an extremely powerful force that, so far, science has not yet found a formal explanation to. It is a force that includes and governs all others, and is even behind every phenomenon operating in the universe and has not yet been identified by us. This universal force is love.
—Albert Einstein (1879–1955)

The First Social Experience

I've been in solitary confinement since lunch, and it's getting dark in my dungeon. The shadows of the tree's branches, swaying in the evening wind in front of the kindergarten, are increasingly difficult to distinguish from the darkness that climbs the stairs from the corridor of the basement, on the lowest step of which I crouch. It's scary. I hope that other things besides the shadows don't start to move. I don't know how long my sentence will last. There are no other children in the house; their voices have long since fallen silent. Everyone has already long since started their journey home. I'm shivering, and I'm afraid. Why I'm sitting here is not quite clear to me. I didn't want to eat the brown-spotted banana. Maybe that was my crime? Or maybe I contradicted the kindergarten teacher too violently when she urged me to do so. What would she have done to me if I had given in and vomited after eating the foul fruit? Having no respect for food was probably what brought me here. It's getting darker and darker. Do I have to spend the whole night here? My eyes fill with tears, and I suppress my need to cry. I sit like this for what feels like an eternity. Finally I hear noises and loud voices, my mother's voice. "Mama!" I shout loudly. "I'm down here!" The door to the cellar is unlocked, and there she is, my rescuer. But her gaze is serious. Now there's more trouble, I think. But no one is angry with me. As I learn later, the kindergarten teacher had forgotten me down there and had gone home.

I almost misspelled the word German word for "dungeon" (*Verlies*). With a "*ß*" at the end instead of a single "S," because for a moment I thought it was the same double S as in "abandoned" (*verlassen*). Because that's exactly how I felt back then. But the word comes from "to lose" or "to lose oneself." This is because long isolation shrinks the hippocampus and causes the soul to suffer. It is true that in our cultural history the heart has repeatedly been mentioned as the place for the seat of our soul; perhaps because we feel how the heartbeat *reflects* our emotional feelings. But the heart only reacts. In my opinion, however, the more logical place for the seat of the soul would be the hippocampus, because it acts or specifically remembers everything we experience, think, and feel, everything that makes up us and our lives. This also applies to us. If we could not perceive ourselves as a continuous self, thanks to hippocampal memory, we would have no answer to the question of who we are—namely social beings who need contact with other living beings like the air we breathe. Without the hippocampus, we would not be able to store emotionally meaningful interactions with other living beings; we would not be able to find soul mates.

Hippocampal memories are the basis of personal relationships, which was impressively demonstrated by their absence in the case of Henry Molaison. The presumably most vital connection is the lifelong love of a mother for her child. It is also based on memories in the first days, weeks, and months after birth, and is therefore dependent on undisturbed growth of the maternal hippocampus. Since this is essential for the child's survival, the evolutionary selection mechanism ensured that a whole armada of hippocampal growth hormones are released during pregnancy, birth, and breastfeeding. For example, the hormones progesterone and estrogen not only maintain pregnancy, they are also potent growth factors for maternal hippocampal neurogenesis. The labor-inducing hormone oxytocin and the milk gland–activating hormone prolactin have the same "additional function."[1]

Thus, the evolutionary biological meaning and purpose of these hormones is not only to ensure a successful pregnancy, a smooth birth, and a nutritious breastfeeding period for the child, but also to strengthen the maternal frontal lobe battery and prepare it to store every little detail of the new relationship.[2] This type of multiple use of hormonal agents is an efficient basic principle of nature that we have already seen with cortisol (eustress improves acute physical performance and at the same time allows the hippocampus to grow). We find them in dozens of other hormones and messengers in all other areas of the formula for a high-capacity frontal lobe battery, such as diet, social and physical activity, and, last but not least, sleep. The more positive and stress-free this new phase of life is felt to be by the mother and the greater detail with which she memorizes all these new experiences, the closer and more stable the relationship with her child becomes in the long term, which in turn is crucial for its survival. No other living creature is dependent so long on maternal care as humans. The reason for this is the enormous learning capacity of our brain, because learning takes a lot of time. So we are very likely to be the most intelligent inhabitants of our planet in our mature state, but as babies we are the most clumsy. Our children need protection and help decades longer than the offspring of most other living beings. Protection from stress during the reproductive phase is therefore important for a stable mother-child relationship. Both prolactin and oxytocin protect the mother from distress and the associated elevated cortisol levels that would damage the frontal lobe battery.[3] Thus, oxytocin not only influences the birth process by regulating labor, it also reduces the mother's birth stress, keeps her blood pressure stable during the birth phase, and improves wound healing afterward. In addition, oxytocin (together with prolactin) activates milk production. During breastfeeding, the mother's oxytocin production increases, as does its concentration in breast milk, and in this way, it also stimulates the growth of the child's brain.[4] Thus, breastfeeding not only strengthens the emotional bond between mother and child;[5] infants who receive breast milk also have better neurological behavioral profiles themselves and are more alert during social interactions. Breastfeeding thus promotes the mother-child bond in both directions.

Breast milk provides the best nutritional composition and the building blocks for the growth of a strong frontal lobe battery, namely through the aquatic omega-3 fatty acids it contains and many other substances, as we will see later. With oxytocin, it also provides a crucial hormonal growth factor that efficiently boosts the production of new nerve cells. This is the GOOD. However, although this type of diet would be natural and species-appropriate, according to the results of a study published in *Lancet* in 2016, only about 37 percent of all infants worldwide were exclusively breastfed in the first six months after birth between 2007 and 2014.[6] A breastfeeding period of two years is considered an ideal and natural period. but here the percentage falls almost to zero. Hardly any mother breastfeeds her baby for two years. This is the BAD and, from WHO's point of view, a huge problem for the majority of humanity. After all, 63 percent of all newborns have poorer physical and, above all, mental development opportunities from the outset because they are breastfed far too little or not at all. Conversely, according to studies by WHO, "breastfed children perform better on intelligence tests, are less overweight or obese, and are less susceptible to diabetes later in life. Women who breastfeed also have a lower risk of developing breast and ovarian cancer."[7] Thus, according to the results of a Brazilian long-term study, which also appeared in *Lancet* in 2015, adults who had been breastfed for at least twelve months as babies had IQs that were four points higher (which indicates better frontal lobe development), were less likely to drop out of school prematurely (which suggests higher psychological resilience and thus the presence of a strong frontal lobe battery), and therefore also had higher salaries than those who had been breastfed for less than one month or not at all after birth.[8]

But it's not just about IQ and career opportunities; the benefits of breastfeeding are even more fundamental. According to further research published in *Lancet* in 2013 and 2016, breastfeeding alone could save 800,000 children worldwide from early death every year.[9] The reason: Breast milk not only provides the best nutrient composition for the child's brain, it is also rich in antibodies and even living immune cells, which, as a passive vaccination, protect the child from deadly infections.

However, breastfeeding also has direct advantages for the mother, which indirectly increases the child's chances of survival. For example, the release of the hormones pro-lactin and oxytocin, which promote hippocampal neurogenesis, protects against post-partum depression. This is a very severe form of depression that some women experi-ence after childbirth and which then also becomes a problem for the baby, as it often suffers from a lack of maternal care.[10] About 11 percent of all women suffer from postpartum depression, about twice as many as a century ago.[11] The cause is often distress due to family burdens or the fear of failing as a mother. This increases the stress hormone level in the direction of chronic distress, which in turn inhibits hip-pocampal growth and thus further reduces mental resilience—a vicious circle. A long breastfeeding period with high prolactin and oxytocin secretions would therefore be a neurobiological opportunity to counteract this danger and break the vicious circle. However, it would be even better to go into pregnancy with a high-capacity frontal lobe battery that provides high psychological resilience—but this is increasingly rare in our modern society.

The UGLY. Based on the clear scientific evidence, a WHO resolution on the promotion of breastfeeding in spring 2018 was to be approved quickly and eas-ily by hundreds of government delegates who met in Geneva for the United Nations–affiliated World Health Assembly. But WHO's plan to use the resolu-tion to restrict inappropriate advertising for substitute foods and unhealthy con-fectionery and thus increase breastfeeding rates and duration worldwide failed. The US threatened Ecuador, which pushed the resolution, with significant eco-nomic and military sanctions if the country voted in favor of the resolution. "We were amazed, horrified and also sad," said Patti Rundall, the political director of the British advocacy group Baby Milk Action. "What happened was tanta-mount to blackmail. The US held the world hostage and tried to overturn nearly forty years of consensus on the best way to protect the health of infants and young children."[12] Was the Trump administration at the time more concerned with the well-being of a seventy-billion-dollar infant formula industry than with the next generation's optimal mental development? Or was it about preserving a future electorate? Unfortunately, this somewhat provocative thought is not so far-fetched, as a lack of breast milk reduces the chance that the next generation will develop into System II thinkers, who are not valued in conservative camps because, compared to System I thinkers, they are less easy to win over as voters and supporters with irrational fear images and populist slogans.

The Kangaroo Mother or the Power of Touch

In addition to breastfeeding, close physical contact is another essential factor for the development of a stable parent-child relationship. The positive effect is most evident when, due to a premature birth, the child's neurological development has to take place outside the uterus much earlier than usual. The so-called kangaroo mother care (KMC) was an idea (re)born in 1978 out of the desperation of the Colombian pediatrician Edgar Rey.[13] He was working at the time in Bogotá at the Instituto Materno Infantil, which served as a maternity hospital for the poorest of the poor and had the largest neonatal ward in the whole of Colombia, with about 30,000 births per year. Due to a lack of money, several premature babies often shared an incubator, which drove up the cross-infection and thus the mortality rate. Rey realized that a natural incubator, that also took in prematurely born offspring, is the kangaroo's pouch. Kangaroo babies are even more immature at birth than most human premature babies. But the direct skin contact in their mother's pouch keeps them optimally warm. Rey put his idea into action: He instructed mothers of premature babies to wear them just as close to the body and bare skin as kangaroos do with their young, and explained to them the importance of breastfeeding. The results were remarkable. The infection and mortality rates decreased, as did the overcrowding in hospitals, as stays became much shorter due to the KMC. The number of premature babies abandoned by their mothers was also reduced because they were able to develop a close bond with their baby, which an incubator did not allow or at least made more difficult.

Positive results were also seen in the intelligence development of KMC children compared to incubator children. This was already the case with one-year-old "kangaroo children" significantly higher, and several long-term studies have shown that this effect was detectable into adulthood.[14] KMC children are less likely to be hyperactive, aggressive, and impulsive later in life as adolescents and adults than premature babies who "traditionally," but completely unnaturally, were brought to maturity with an incubator.[15] Their attention span is also significantly longer. However, the positive effect of KMC in premature babies is also can be achieved by Kangaroo Father Care (KFC), which states that direct skin contact is of independent importance for the child's brain maturation, regardless of breastfeeding.[16] There are also indications that physical contact in KFC also has an effect on the father. In this way, it strengthens long-term cohesion, because KFC fathers leave their families less often and thus more often become grandfathers who are firmly integrated.

Should you let a child cry? This is a question that science actually asked itself and was seriously investigated in the 1970s. A pioneer in this field was the American Canadian developmental psychologist Mary Ainsworth (1913–1999). She found out what should be clear to everyone today, namely that infants whose mothers ignored their crying (mostly for "educational" reasons) deepened their sadness and cried more often, while those whose mothers responded to the crying with affection cried less often. The mother's absence, despite clear crying signals, also impaired the later exploratory behavior of the children, their hippocampal curiosity, while the presence of the mother promoted it.[17] Basic trust is the foundation for this. The prerequisite for basic trust is reliable, loving, and caring attention from parents (or other caregivers). Basic trust is synonymous with an emotional and social sense of security and thus enables a fearless exploration of the environment and a trusting confrontation with the social environment. Basic trust allows you to love because you feel loved, and to trust because you could trust as an infant. Through the attention paid to the child, broader communicative skills in facial expressions and vocalizations also developed. Further studies confirmed that children who are allowed to cry are significantly more likely to have mood and behavioral problems.[18] The attention that an infant receives in the first months of its life influences its brain development and thus its entire life.

It is obvious that the principle of KMC or KFC contributes to the better brain development of *all* newborns, including those born on the calculated date of birth. Finally, direct skin contact with the mother or other caregivers in the first few months of life was completely natural for a long time in human evolutionary history.[19] This can still be observed today in indigenous peoples or African farming families.[20] Studies show that even completely healthy babies born at the calculated due date benefit from early skin-to-skin contact (SSC, the scientific term), because it leads to a faster maturation of the genetic program.[21] Such changes in the genetic material, referred to as epigenetic, regulate which genes become active at what time. It turns out that the genetic program in those children who had comparatively little skin contact with their parents, and thus suffered more from newborn stress, matured more epigenetically slowly.

Behavior is inherited epigenetically. At the beginning of the nineteenth century, French zoologist Jean-Baptiste de Lamarck (1744–1829) put forward the thesis that organisms can inherit characteristics that they acquire during life. For almost two centuries, his revolutionary idea was overshadowed by the theory of evolution of Charles Darwin (1809–1892), who postulated that newly hereditary traits only arise through random mutations in the formation of germ cells (egg and sperm). But de Lamarck was also to be proven right. For example, young mice not only improve their own learning ability when they are mentally stimulated in a stimulating environment, but also pass this ability on to their offspring by means of the epigenetic imprint (epigenesis). In contrast to the offspring of mice, which had to eke out their lives in a laboratory cage that was not very stimulating, the offspring of such privileged mice were even able to overcome a congenital genetic (i.e., purely Darwinian inherited) defect by means of epigenesis, which would normally have limited their ability to learn. Epigenetics, i.e. inherited experience, beats genetics.[22]

However, epigenetics works in both directions. Further research has shown that in adults who received less physical affection as children, the genes for the development of oxytocin receptors and for the brain-derived neurotrophic factor (BDNF, a factor released in the brain that activates brain growth and is perhaps its most important growth factor) became less active due to epigenetic changes.[23] However, this result was not surprising, since it had previously been shown that the oxytocin receptor is epigenetically inhibited by chronic distress.[24] As a result of this epigenetic imprinting of its receptor, oxytocin can no longer work efficiently, with all the negative consequences for mental, emotional, and social development.

A team of Canadian researchers examined the genetic makeup of thirty-six individuals who died at the young age of around thirty-five.[25] One-third of them had committed suicide due to childhood abuse, another third had taken their own lives due to other mental health issues, but—like the last third—without such a negative childhood experience(s). The last third did not die by suicide, but by accidents. However, only the abuse group had the gene that encodes the hippocampal cortisol receptor chronically inactive due to epigenetic imprinting. Without the sensor for circulating cortisol, the hippocampus could no longer react adequately to stress; even eustress became distress. This was a clear indication that the tendency to severe depression, which, as in this study, can even end in suicide, is not caused by the genes themselves, but by their epigenetic imprint. In the meantime, 365 more "hippocampal" genes that may have been altered by early negative experiences with the environment have been identified.

On the other hand, if one has the good fortune to grow up in an environment that promotes mental activity, it is mainly those genes that are responsible for high synaptic plasticity (a high learning ability) that are read.[26] This type of genetic fine-tuning even works (epigenetically) across generations, as has been discovered in animal experiments.[27] The quality of human episodic memory and thus frontal lobe battery performance is also shaped epigenetically. Epigenesis thus influences hippocampal neurogenesis and thus also psychological resilience as well as the individual risk of developing depression and, in the long term, Alzheimer's disease. According to New York psychologist Catherine Monk, this type of imprinting begins very early: "The starting line for psychological development shifts well before birth."[28] Conversely, maternal love, a close mother-child relationship, increases the capacity of the child's frontal lobe battery. The stronger the mother-child bond, the higher its capacity and performance.[29]

> **The UGLY.** Children and adolescents who have been beaten have a comparatively small frontal lobe[30] and a weaker frontal lobe battery as adults.[31] This has mental and psychological consequences, such as the development of depression or a lack of self-esteem. Nevertheless, humiliation of children and beatings are also commonplace in Germany, and according to crime statistics, the number of victims is increasing, especially because of the COVID measures.[32] Violence against children occurs in all social classes.[33] France only introduced a "right to non-violent education" at the end of 2018; Sweden has had such a law at least since 1979. In Germany, parents and teachers have only been prohibited from corporal punishment since 2000. In Germany, in contrast to France, this has at least criminal consequences—if a complaint is filed, which happens far too rarely.

Hormone of Fidelity

Oxytocin, like hardly any other hormone, highly efficiently stimulates frontal lobe battery growth and thus creates new nerve cells that ensure the emotional connection between the hippocampi of two individuals. Based on a large number of research results in this regard, oxytocin is probably rightly considered a pleasure, cuddle, and ultimately a loyalty hormone. Oxytocin not only increases the emotional bond between the mother and her newborn child, but also strengthens the long-term pair bond within the context of sexual experiences.[34] Thus, a strong hippocampal neurogenesis stimulated by oxytocin ensures that, as a rule, no one forgets their first kiss—even into old age. The oxytocin system promotes monogamous fidelity and thus increases reproductive success because it is much more likely that both parents will care for their offspring's well-being together and in the long term. This "biology of fidelity" was studied in great detail on prairie voles. The oxytocin system also makes animals monogamous by nature. However, if this is inactive due to a genetic change (mutation), as in the case of the mountain voles related to them, the result is a rather promiscuous lifestyle—no trace of loyalty. However, if their oxytocin system is reactivated by artificially administering the hormone, they also become monogamous.[35] This also seems to work in us humans.[36] For example, married men, in contrast to single men, kept a greater distance from attractive women when they had previously been given a pinch of oxytocin in their noses, unnoticed. For success in the evolutionary biological sense, it is not enough to simply produce offspring—the oxytocin-induced pair bond also improves the conditions for the children to mature healthily and have a higher chance of having children of their own. Otherwise, the transmission to the next generation and thus the existence of one's own genetic material ends.

In humans, the ability for lifelong hippocampal neurogenesis became the evolutionary biological means to an end: The older generation's empirical knowledge also increases the chances of success of the second generation.[37] In addition to many other hormones, oxytocin ensures this. However, it not only stimulates hippocampus growth, it also has a transgenerational effect: it not only influences one's own social behavior, but also shapes the next generation's.

However, oxytocin also ensures close, long-term bonds beyond one's own family ties, because it also shapes and strengthens friendships. The oxytocin level in the blood corresponds to the level of trust.[38] It is therefore also considered a hormone of social cohesion. Finally, the oxytocin-stimulated hippocampal neurogenesis is the remembrance of rituals and traditions experienced collectively, which in turn influences social behavior. Such rituals and traditions that shape the community ultimately make oxytocin a true "culture hormone" and at the same time make the human being a herd animal.

Oxytocin, as part of our genetic program that emotionally attunes us to group-conforming behavior, was decisive in the coevolution of intelligence, social behavior, and human language.[39] The sociocultural pressure ensured that those who were most likely to pass on their genetic material were those who knew best how to use the complex social structures in the community for their own benefit, who had a certain degree of social intelligence. This included being able to remember how other people behave and, above all, feeling what they expect from you. However, belonging to a group also has disadvantages. On the one hand, you feel comfortable in it because it offers security and strengthens the individual; on the other hand, groups are usually defined by characteristics that distinguish them from others. The consequence of herd behavior is often blind obedience, following hidden rules of interaction that are harmful to oneself or others.[40] The effects become particularly clear when blind trust in one's own group leads to unjustified mistrust of foreign groups or ethnicities, or even fuels xenophobia.

Blind trust despite a clear view or conformity at any price. In order to test the power of the group over the individual, American social psychologist Solomon Asch conducted one of the best-known psychological experiments in the middle of the last century.[41] A large group of students was to jointly estimate the length of a line compared to the length of other lines on a card. However, it was not a matter of testing visual accuracy, because Asch instructed all but one or very few students to commit to an obviously incorrect estimate before the test. Those study participants who did not know about this secret agreement were the actual test group. They believed that the other students would decide freely like them. Nevertheless—or precisely because of this—they also decided, obeying the majority, mostly for the wrong assessment and thus against their common sense. This "blind trust" in the obviously wrong was all the more likely the smaller the test group was compared to the majority. Asch was thus able to show that individuals or a few, under peer pressure, will decide against the subjectively correct feeling and for an objectively wrong result.

In groups, external beliefs usually weigh more than internal ones; this applies to supporters of religious and ethnic communities as well as to supporters of a football club. On the one hand, as sociologist Eckart Voland described it, "as in football, you can indulge in the regression to pack formation—and all this in a pleasant party mood."[42] On the other hand, those who do not participate or cannot participate because of skin color or religious beliefs, that is, do not behave uniformly (and are usually not even outwardly "uniform"), do not belong to the group and are isolated, which can be dangerous for those affected. Racial hatred is a possible consequence, but also rejection and fighting against anyone who thinks differently. In such sworn groups with a clear image of the enemy, thinking or not thinking on the basis of System I predominates, despite the pleasant release of oxytocin. Although this promotes the togetherness, it is essentially based on stereotypical slogans and prejudices.

Mirror Neurons and the Baby Schema

In addition to breastfeeding and direct skin contact, eye contact as well as facial expressions and gestures are of enormous importance for the development of the infant's social skills. Since a baby is not yet proficient in human language, their brain development, from an evolutionary biological point of view, is in a certain way in an early phase of becoming human. Even our prehuman ancestors had not yet mastered complex language, and the essential part of communication took place via body signals, as with primates. But even this ability enabled them to recognize dangers, such as sensing the mood of another person: Are they well disposed toward me, or are they planning evil? Recognizing this was essential for survival and can still be observed in the animal kingdom, especially in our closest relatives, the primates.[43]

Facial expressions and gestures still function like a language in their own right. To learn them, we mirror the people around us. Mirror neurons are already present at birth for this purpose. They allow the infant, just a few days after birth, to mirror its caregiver, such as smiling when it is looked at with a smile or yawning when the other person yawns.[44] This mirroring is a basic newborn emotional need because only in this way can it gradually learn to sense other people's feelings.

To yawn or not to yawn, that is the question here. Anyone who has ever seen a crime thriller with serial killers knows that psychopaths are selfish, manipulative, impulsive, fearless, callous, and often dominant and especially without empathy. They cannot mirror their victims' emotions. This group of people is therefore also "immune" to otherwise contagious yawning. People on the autism spectrum show this phenomenon just as often[45]—but without psychopathic tendencies. Their "immunity" is due to the fact that they usually avoid looking their counterpart in the face because they have problems interpreting facial expressions.[46]

Childlike facial features such as large, round eyes and a round face with a snub nose awaken in people the almost compelling need to look into the face of a child (or another, usually young creature) and smile at it—but also to mother it, protect it, and take care of its well-being. As a reward, you get a release of hippocampal growth hormones such as oxytocin and dopamine, which provide mental well-being. This reward scheme could be the reason why evolution has produced the infant schema.[47] Interestingly, it works for humans not only with our own offspring, but also with the observation of animals that have these characteristics. For example, a dog owner releases oxytocin when looking into the eyes of his four-legged friend—but vice versa also in the dog, which speaks for a coevolution of the oxytocin system.[48]

The development of mirror neurons, which are so important for social competence and empathy, reaches its peak in the third and fourth year of life, when children begin to distinguish themselves from other people and learn social skills. "Learning" is the right term for this, because this does not happen automatically, but like many other brain functions requires training. However, the social training of mirror neurons is now in danger. Of course, there are many reasons why more and more mothers are looking

Figure 16: Competing child schemes. Childish schemes, such as those of the Japanese emojis, products of our cultural evolution, have become highly attractive to many parents.

at their mobile phones more and more often than at their child's face; perhaps they are more attracted to the friendly winking emojis in cyberchat as an expression of successful social interaction, which presumably do not correspond to the childish scheme purely by chance.[49]

According to the results of a 2015 study, young adults looked at their smartphones an average of eighty-five times a day, spending a total of about five hours a day.[50] This leads to disruptions in the relationship with family members present, especially in the youngest, who depend on parental eye contact for their mental and, above all, emotional development. The *Süddeutsche Zeitung* reports under the headline "Turn off the mobile phone": "Instead of dealing with their children, many parents distract themselves with their smartphones. This affects the little ones and could promote behavioral problems."[51] That applies, above all, to their social skills. If parents don't pay attention to their media consumption and reflect to their children that the screen is more important, they will "get the receipt five years later,"[52] says Sabina Pauen, developmental psychologist at the University of Heidelberg. Attachment disorders, a lack of empathy, and concentration problems could be the result. Till Reckert, pediatrician and media officer at the Professional Association of Pediatricians and Adolescents (*Berufsverband der Kinder und Jugendärzte*, BVKJ), also warns: "The more the mobile phone has 'grown,' the more likely it is to hinder the necessary presence for the children's education of the next generation." The attachment disorders caused by smartphones are already causing problems with the little ones falling asleep. But that's not all. At the age of four months, "smartphone children" learn to avoid glances—completely against nature. "It's unpleasant when the mother doesn't look back, so I'd rather not look," explains Pauen. "Even very small children then resign." Without training in social skills, which include a correct interpretation of facial expressions and gestures, children are thus on an "autistic path," which could explain why this disease is on the rise more than any other. In the middle of the last century, only one of about two thousand children suffered from autism, which was 0.05 percent. The rate is now over forty children per two thousand, which is over 2 percent.[53] No other disease of the child's mind, not even ADHD, has increased so rapidly.

Social Isolation

In prehistoric times, being socially isolated meant certain death. Our ancestors had to protect themselves against predators around the clock and defend their livelihood, the seafood-rich coastal strip, against invading groups, which could only succeed in community. Even those who were sick could only survive with the help of their clan.

Avoiding these dangers fundamentally shaped human behavior on a genetic and thus also on a neuronal level.[54] We became beings who need a family or a social environment throughout our lives to feel safe and secure. As a result, we like to have parties and celebrations, play sports, or do other things with other people. Such social activities are crucial for the formation and survival of new nerve cells in the hippocampus. Our social life increases our mental resilience and thus also protects us from stress-related diseases.

In particular, the first years of child development are formative for social behavior. Deficits in the social sphere thus have long-term consequences, as shown by studies on adolescents who grew up as young children in Eastern European orphanages—usually without emotional and physical affection.[55] Even after successful adoption, their oxytocin release remained far below that of children who grew up in intact families from the beginning.[56] The consequence of social isolation is a long-term loss not only of trust in others, but also of self-confidence and self-esteem, and not least of empathy—clear indications of a disturbed development of the frontal lobe *and* its battery. Other symptoms of social isolation include psychiatric and neurological disorders such as anxiety, depression, schizophrenia, memory loss, and even epilepsy—these were also confirmed in animal experiments.[57]

Being socially isolated is one of the worst things that can happen to a person, but you can also feel lonely in a group. The depression researcher Martin Hautzinger from Tübingen provides the following explanation: "Being alone is a social phenomenon: You are not together with others. Loneliness, on the other hand, is a psychological phenomenon. You can also feel lonely among hundreds of people."[58] Whether perceived or actual physical isolation, the person suffers, stress hormone levels are chronically elevated, and the frontal lobe battery loses capacity. But even without an increase in cortisol, social isolation selectively degenerates the hippocampus because it reduces the release of BDNF, one of the most important and potent growth hormones in the brain. Due to its reduction and many other such mechanisms, the social memory in particular loses its efficiency. Experiments on young primates confirmed the finding that social deprivation or loneliness, especially during the transition from puberty to adulthood, which is so sensitive to brain development, leads to enormous stress. This is the stage of life in which a primate usually seeks his life partner, and loneliness is not helpful. Permanently anxious behavior can be the result, which in turn inhibits adult hippocampal neurogenesis, impairs mental resilience, and thus—like a vicious circle—causes long-term persistent stress states.[59] Accordingly, Angelika Erhardt from the Max Planck Institute for Psychiatry in Munich concludes: "There are obviously particularly vulnerable phases in early childhood or later in puberty in which stress has significantly stronger effects at the molecular level."[60] In addition, loneliness has a repellent effect on others, which creates a frightening vicious circle: When an individual is perceived as lonely, other people (unconsciously) avoid interaction. The stigma of loneliness leads to a self-reinforcing cycle of social isolation.

On the Minimum of the Social

The scope of social activities is subject to individual needs, but also to the laws of minimum and maximum. For example, complete isolation would be an extreme minimum that hardly anyone can cope with. Especially in older age, social isolation increases the risk of hippocampal dementia, especially Alzheimer's. The cause of hippocampal neurodegeneration is, on the one hand, loneliness stress and, on the other hand, a lack of social activation of hormonal growth factors. "Loneliness accelerates mental decline, as studies from all over the world show."[61] Lisa Berkman, head of the Department of Society, Human Development and Health at Harvard University, followed over 16,000 retirees over a long period as part of the Harvard Study.[62] In a press report on the study, she summarized her findings: "We know from previous studies that people with many social connections have a lower mortality rate. We now have overwhelming evidence that close social networks can help prevent memory decline."[63] But this is exactly where today's society faces a growing problem: we live in an era of chronic loneliness. With the cultural trend toward individualization and the resulting increase in the number of single households, the problem of sickening loneliness is increasing, not only in Germany, and this also applies almost every fourth young person. Experts speak of a "modern epidemic."[64] According to a response by the German government to an inquiry by the FDP (Free Democratic Party), the loneliness rate in Germany also increased by around 15 percent just between 2011 and 2017 among forty-five- to eighty-four-year-olds, and in some age groups by as much as 59 percent.[65] In its report, the German government cites, among other things, a study by the market research institute Splendid Research from 2017. According to this, 12 percent of Germans often or constantly feel lonely. This is particularly common for people in their mid-thirties, 18 percent of whom are affected. Referring to scientific research, the federal government comes to the conclusion that "social isolation in particular has an unfavorable influence on the occurrence and course of chronic diseases." This was shown in connection with high blood pressure and other risk factors for cardiovascular diseases, but also chronic obstructive pulmonary diseases as well as mental illnesses (depression, anxiety disorder, suicide rate) and dementia. The effects of being alone on mortality are clear from the results of a study by Brigham Young University. According to the study, "loneliness is as harmful as smoking or obesity in terms of all-cause mortality."[66]

Ministry of Loneliness. Not because the British are isolating themselves from the EU, but because loneliness is a huge problem in the island state,[67] the government created a corresponding ministry in 2018. According to Mark Robinson, chairman of Age UK, Britain's largest charity for older people, loneliness kills: "It has been proven that it is worse than smoking 15 cigarettes a day."[68]

The COVID-19 measures had made this development worse. Current studies indicate that the isolation measures against the coronavirus pandemic have profound psychological and social effects, as they reinforce this cultural trend toward increasing loneliness.[69] A massive increase in depression and dementia is therefore to be expected in the short but also long term due to the COVID measures.[70] System II thinking is also likely to have suffered as a result of the measures. At the end of the psychological consequences of social distancing and social isolation, which impair the functionality of the frontal lobe battery and reduce psychological resilience, there is an increased suicide rate.[71] Children too, and above all, suffer. According to Dr. Thomas Fischbach, president of the Professional Association of Pediatricians and Adolescents, mental and physical side effects are considered to be "collateral health damage" from the lockdown that can be clearly seen due to daycare and school closures.[72]

The UGLY. We live in a culture of increasing individualization and are increasingly becoming "lone wolves." With the loss of extended families, the social support that provided our ancestors with stability and security is missing. As a result, the feeling of loneliness is growing in modern industrial societies, with many health consequences, not least with a loss of frontal lobe battery capacity.[73] In order to survive, we believe we have to become part of a larger collective, and we adapt. The advertising industry tells us what we have to do, how we have to dress, and what technical gadgets we have to have in order to be "in" or at least to be able to feel "in." "So we buy things that we don't need," as an unknown author once aptly put it, "with money we don't have, to impress people we don't like." Here, the fear of social isolation meets a social pressure that is hard to withstand with a weakening frontal lobe battery. You need a lot of self-confidence and a strong sense of self-worth to go your own way.

From the Maximum of the Social

In the long evolutionary history of humans, the number of social connections that were intensively cultivated was manageable. You had a close relationship with your family and perhaps one or two with members of your own or neighboring clans. Basically, you were part of a close-knit and relatively stable community. Today, however, people are becoming increasingly individualized, the "I" stands above the "we," and the number of close social contacts that a member of modern society maintains is decreasing. In order to still be "in," to feel part of a close-knit community and to always know what is "possible," social media presence is increasingly becoming a substitute and at the same time a must. We can now be constantly in touch with family, friends, colleagues, and all networks around the world via mobile devices. This sounds good at first, but the inexorable flood of information as a result of this modern type of "social activity" can

quickly overwhelm us. Even though the human brain is extremely powerful and receptive, this still pushes our archaic brain to its limits.

The constant accessibility that social media makes possible and that its users therefore also demand causes stress and reduces our "own time," the time we spend and need
for ourselves (this includes in particular, but not only, sleep). Social activity, on the
other hand, "foreign time" that one takes for others, is taking up more and more space
through digital communication. Both time spent with others and time spent alone
are vital, but at the same time they are in direct competition with each other, which is
expressed by their diametrically opposed position (expressed as "Social Activity" and
"Sleep") in the formula for a strong frontal lobe battery (see Figure 14 in Chapter 7).
Too much foreign time, spent in cyberspace, inevitably causes an information overload
of the frontal lobe battery and at the same time also creates a lack of own time, resulting
in sleep and concentration disorders as well as mental exhaustion and even burnout and
depression.[74] This can be observed especially among the younger generation. Studies
show that academic performance decreases with increasing amounts of social media
activity.[75] Even before entering kindergarten, depending on the time toddlers spend
in front of screens, detrimental structural abnormalities in the brain are detectable.[76]
Many other studies have shown that high screen time is related to the inability of
children to pay attention and think clearly, but also promotes poor eating habits and
exacerbates sleep and behavioral problems. A delay in language development, impaired
executive function, and a disorder of the parent-child bond are also typical.

We need social contact, but quantity, as modern media allows, does not replace
quality. On the contrary: one promotes and the other inhibits spiritual development.
"Internet friends" are no substitute for real friends. A chat over text is no substitute for
a real, lively counterpart. In addition, negative things are easier to write than to say to
someone's face, and last but not least, people usually only post what they believe others
will admire (example beauty filter), but in turn, these admiration posts usually only
cause envy and dissatisfaction with dramatic consequences for mental (hippocampal)
health.

CHAPTER 10

WHY WE SLEEP

Let sleep be the daily bread of your soul.
—Carl Ludwig Schleich (1859–1922)

The early bird can take me . . .
—(my coffee cup, author unknown)

Sleep Drunkenness (Fog from Sleep)

"You're not a doctor," the elderly lady shrieks at me and resists while I try to reinsert the catheter for the infusion.

In a state of mental confusion, the patient had recently torn it out in the middle of the night. Whether it's due to the sweat on my forehead or my young face, it's hard to say; in any case it's not hidden from her despite her confusion that I'm not yet licensed and that I'm in my clinical year (PJ), a few months before the final medical examination. The main tasks of a "PJler" are to draw blood and insert IVs, and actually, after almost a year of doing this, I'm quite well practiced. But the cannula just doesn't want to find its way into the woman's arm vein.

This may be due to her verbal abuse and the fact that the patient resists violently, so the night nurse has to hold her down. But certainly also because I was just in a deep sleep and now, still drowsy, I have to overcome a technical difficulty, which I just don't want to succeed in.

That event was responsible for my professional career. By then I had kept both options open for later in order to be able to work either as a doctor or as a scientist. Although a torn-out infusion like this can be renewed and does not necessarily determine the life of a patient, not being fully in control of my mental powers due to lack of sleep frightened me. My fear that I would make a serious mistake as a responsible doctor under such conditions tilted my career plan toward medical research. Here I hoped to be able to

117

contribute to people's well-being in a different way. At least that way I would no longer run the risk of harming a patient due to overtiredness.

Sleep drunkenness is the state of psychomotor slowing down with decreased cognitive performance after waking up. You are disoriented. Perception and processing of external stimuli and the reaction to them are limited. Actions appear purposeful but are often incorrect. The word *drunkenness* in connection with sleep refers to a state that is in fact very similar to that under the influence of alcohol. The term could therefore not only be used for the mental state shortly after being woken up from a deep sleep, but also for when one has been awake for too long and urgently needs to sleep. After just seventeen hours without sleep, your performance is as impaired as if you had a blood alcohol count of .05 percent; after nineteen hours, it's like .10 percent.[1] The fact that you have to make potentially life-or-death decisions with such values and the resulting reduced reaction times, for which you would have your driver's license revoked in most countries, is irresponsible in my opinion.[2]

Glutamate provides a scientific explanation for the "drunkenness" caused by lack of sleep. As already discussed, the neurotransmitter activates new experiences or thoughts in the hippocampal synapses. As a result, they change their conductivity, which corresponds to the storage of these experiences or thoughts. To ensure that new memories are actually stored via free synapses and that no previous ones are overwritten and deleted, β-amyloid released by glutamate inhibits the further release of glutamate at synapses that are already in use, a form of negative feedback that secures our memories. However, with each new memory, slightly more β-amyloid accumulates in the hippocampus, and at the same time, the number of free synapses decreases; our mental energy "depletes." Therefore, one of the important tasks of sleep is to break down β-amyloid after memories have been "uploaded to the neocortical hard drive" in slow-wave sleep (see Chapter 4). This is the only way the hippocampus is ready to absorb again the next morning.

Like β-amyloid, alcohol also inhibits the hippocampal glutamate system, consisting of the neurotransmitter itself and its synaptic receptors, not synapse-specifically, but on a broad level.[3] Sleep deprivation and alcohol thus have a similar effect on the hippocampal memory system; both deplete our ego, which explains the mostly stereotypical thinking under the influence of alcohol, but also why people are more inclined to tell the truth when they are drunk. Ultimately, successful cheating or lying requires a functioning System II. Lies have the proverbial short legs because it is exhausting to keep them going.

The Frontal Lobe Battery Only Grows During Deep Sleep

As we already know, however, nocturnal deep sleep is not only for uploading new memories from the hippocampus to the neocortex. The subsequent degradation of ß-amyloid is necessary, and hippocampal neurogenesis can only take place if the hippocampus is "offline" and not busy saving new memories. The optimal hormonal conditions are created before and during sleep for the production of new nerve cells, as Figure 17 illustrates. For example, even before falling asleep, the stress hormone cortisol is downregulated to such an extent that it does not inhibit neurogenesis. On the other hand, melatonin is upregulated at the beginning of night sleep to initiate deep sleep and activate hippocampal neurogenesis, and growth hormone, which is also released at the beginning of night sleep, is a potent activator of hippocampal neurogenesis.[4]

Incidentally, this constellation—cortisol low, growth hormone high—is also the reason why wounds heal better during sleep, because every organic repair requires new cells. In relation to neuronal nerve cell production, one can now well imagine what happens when sleep is interrupted and one is suddenly confronted with a stressful situation or negative thoughts. The resulting change from a low or regenerative cortisol/growth hormone ratio to a high one stops neurogenesis and even threatens neurodegeneration.[5] Our mental abilities suffer; deep sleep is a basic prerequisite for our brain to become receptive again and for us to be able to use System II.[6] Even one night of poor sleep disrupts memory and mental performance. The new day begins in a still ego-depleted state. But chronic sleep disorders in particular, especially those with frequently interrupted deep sleep, are harmful to our frontal lobe battery.[7] A chronic frontal lobe weakness due to a low-capacity frontal lobe battery, the development of depression, and Alzheimer's disease, which, due to its origin, can certainly be referred to as hippocampal dementia, are all threats. However, the child's brain is particularly sensitive to a lack of sleep. Not only the hippocampus of children grows during sleep, but also their entire brain.[8] This represents an enormous individual but also a social problem, because very few children now have enough bedtime available.

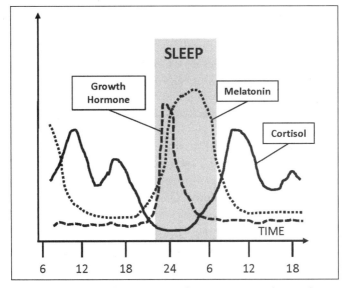

Figure 17: Hormonal prerequisites for neurogenesis during sleep.

Overtired Society

Insufficient sleep is a huge problem of the modern age. About 100 years ago, for example, adults rested an average of nine hours, which is much closer to our natural need for sleep than the seven hours or so we allow ourselves on average today. Since the light bulb's invention in 1913, we have lost around two hours of sleep per night, although Thomas Edison, in keeping with the zeitgeist of the time, is said to have said that people sleep 100 percent more than necessary. However, this grotesque attitude has not changed much to this day, despite the fact that we know about the essential importance of sleep. According to psychologist Bärbel Kerber, sleep degenerates "the residual size in our modern lives."[9] In *Psychology Today*, Kerber quotes the sleep expert Stanley Coren: "What people today demand from their bodies and minds contradicts their biological nature." The consequences are serious. According to the DAK report "Germany Sleeps Poorly—An Underestimated Problem" (*Deutschland schläft schlecht ein unterschätztes Problem*), around 80 percent of all employees were affected by sleep disorders in 2016. That's around 34 million people. If you compare this with the figures from 2009, this corresponds to an increase of 60 percent.[10] Almost half of all employees are exhausted at work and about a third are almost constantly exhausted. Chronic ego depletion, solely due to inadequate regeneration of the frontal lobe battery, is a mass phenomenon.

How much sleep do people need? As is well known, the cradle of humanity was in Africa, near the equator. Day and night are about the same length there all year round. In humans, as diurnal creatures, evolution adapted its genetic program to this given rhythm. Sleep researchers have actually determined that when not influenced by an external timer, humans naturally settle into a sleep time of between nine and ten hours.[11] But these are only average values. You can easily estimate your individual sleep needs. If it takes you a long time to wake up in the morning and requires an alarm clock, which might wake you from a dream, you probably needed some more sleep. A high need for caffeine also signals that there is a lack of sleep.

Insufficient sleep at night is the cause of daytime tiredness and exhaustion, depressive moods, and many other health problems. A habitually short sleep duration is associated with obesity, type 2 diabetes, hypertension, cardiovascular diseases, and depression—even in adolescents.[12] One example is children who have respiratory problems and therefore tend to snore. You have restless, frequently interrupted sleep. This has serious consequences: Their nerve cell production is disrupted, and the nerve cell layer in the frontal lobe in particular is thinned out compared to children without sleep problems.[13] As a result, they often have difficulty concentrating, lack of impulse control, and learning problems. However, these children are just the tip of an iceberg, because lack of sleep is now common among many children and especially young people—even those without respiratory problems. The health consequences of chronic lack of sleep are considerable, but of all the organs, the brain suffers the most.

This was also impressively demonstrated in young rats, which were allowed far too little sleep for a few weeks with only four hours per night (the rodents are nocturnal and sleep seventeen to eighteen hours a day). The result was a massive shrinkage of their hippocampi.[14] Not only the lack of sleep in itself, but also all the aforementioned secondary diseases hinder the growth of the frontal lobe and its battery, thus increasingly providing for a new generation of people whose mental performance and mental abilities on average fall further and further behind their potential.

Chronobiologist Jürgen Zulley summarizes the results of today's research succinctly and unvarnished: "Too little sleep makes you fat, stupid and sick"[15]—but also unattractive and unpopular. This is the result of a study that also provides an evolutionary biological explanation for it. Since we can't tell at first glance whether someone is simply sleep-deprived or perhaps sick and therefore potentially contagious, we subconsciously keep our distance to be on the safe side.[16] "The less sleep you get, the less you want to interact with others. In return, you have a social impact, which amplifies the social consequences of sleep loss," explains brain researcher Matthew Walker regarding the findings of another study titled "Sleep loss causes social withdrawal and loneliness," which he published with his colleague Eti Ben Simon at the University of California, Berkeley.[17] Lack of sleep thus has social effects, and this in both directions: Not only are you more likely to be avoided if you don't get enough sleep, you are also less social due to ego depletion, and your executive functions suffer.[18]

The negative effects of chronic sleep deprivation on executive functions have even been noted in children aged three to five: Those who get too little sleep in these early years are worse at organizing and planning at the age of seven.[19] Children who lack sleep are more inattentive as well as hyperactive, thus showing the basic symptoms of ADHD. This significantly affects school performance and also causes social problems. If children and young people are given stimulants due to an ADHD diagnosis, drugs with an active ingredient such as methylphenidate (known as Ritalin), they sleep even worse, which not only does not solve the basic problem, but exacerbates it. This is the core finding of a meta-study published in the *Journal of Pediatrics*. "Stimulants led to longer sleep latency [later sleep onset], poorer sleep efficiency, and shorter sleep duration," summarizes Katherine M. Kidwell in her analysis.[20] Chronic sleep deprivation as a possible cause of ADHD and a negative effect of drug "therapy," namely sleep deprivation caused by stimulants, reinforce each other. This is a vicious circle that damages the children's mental development in the long term. The only right solution would be species-appropriate sleep, but very few people get that.

"Pep pills" for kids and managers. To enable the Blitzkrieg, the Nazis introduced methamphetamine, a stimulant that is in circulation today as crystal meth. Dubbed "tank chocolate" or "aviator marzipan," the soldiers used this to pep themselves up and feel less fatigue, exhaustion, fear, and hunger. Chocolates laced with methylamphetamine were even available as so-called "housewives' chocolate."[21] The ADHD therapeutic drug methylphenidate, marketed as Ritalin, Medikinet, or Concerta, is structurally similar to methylamphetamine and is traded in the drug scene as a replacement drug for "speed" (amphetamine).

Sleep Hygiene

Maintaining good sleep hygiene means doing everything essential that improves our sleep quality (Law of the Minimum) while avoiding everything that worsens it (Law of the Maximum). The formula for a strong frontal lobe battery refers to all areas of life that need to be taken into account:

Meaning in life: Everything that fills our lives with meaning keeps us busy. But the associated challenges should not prevent us from sleeping. Instead of brooding ineffectively at night with a (still) empty frontal lobe battery, it is better to deal with the topic at hand the next morning. It can be helpful in two ways to write down all the aspects that come to mind before going to sleep: For one thing, we don't have to keep these thoughts circling in the frontal lobe just for fear of forgetting them. On the other hand, our brain will write down everything that we consider important before we sleep related to previous experiences while dreaming and will possibly find new solutions that often come to mind "surprisingly" when we wake up in the morning—as if they had come out of nowhere. So you just have to trust sleep as a problem solver, then you can sleep better. Unfortunately, this is easier said than done, because many people lack this trust, and they brood instead of resting.

From trauma to nightmare. According to modern linguistic research, the word *trauma* is derived from the Germanic *drauma* for wound; the origin of *dream* is the Greek word for *wound*. If trauma prevents us from dreaming healthily, psychological trauma therapy is advisable. Without healing or processing the emotional wound, falling asleep and staying asleep usually remain disrupted in the long term.

Social activity: Our social activities during the day have a significant influence on how we sleep. These now increasingly include social "networking," dealing with digital media. The so-called "digital detox" is therefore becoming an increasingly important part of our sleep hygiene, and this also applies to children. They increasingly use digital technologies such as cell phones, computers, and television late in the evening, with a significant negative impact on the quantity and quality of their sleep.[22]

According to the prevention radar of the German Employee Health Insurance Fund (DAK), almost a third of all students suffered from serious sleep problems in 2018 due to long screen time.[23] Half of all students surveyed complained of fatigue and felt exhausted during the day. Teens who reported spending more than four hours a day in front of screens slept about 7.3 hours on average, around 1.6 hours less than those who used smartphones, tablets, and similar devices for less than an hour a day. Andreas Storm, CEO of DAK, fittingly remarked: "Students take care of full batteries in their smartphones at night, but they no longer sufficiently recharge their own batteries."[24]

There are computer programs that reduce sleep-disturbing blue light in the evening. They then automatically switch to night mode.[25] However, cell phones not only emit blue light, but also microwaves, which activate the stress system and can even cause inflammatory reactions in the brain. This increases the rate of neurodegeneration in the hippocampus, while at the same time inhibiting neurogenesis and presumably increasing the risk of cancer.[26] You should therefore always use your mobile phone with a headset and as rarely and briefly as possible, and it is better to have longer conversations via an analog phone (if you still have one).

Nutrition: A few hours before going to bed you shouldn't eat anything that puts a strain on your digestion and therefore prevents you from sleeping. If you feel hungry late at night, a handful of nuts is a good choice, because it's not a good idea to go to bed hungry.

Alcohol, although it is often drunk to "switch off" and can make us tired, actually makes it harder for us to sleep through the night. It also directly inhibits hippocampus growth. Ideally, you should avoid alcohol completely (with exceptions), because according to the latest research, every sip is a sip too many. Instead, calming herbal teas are recommended. These include valerian tea (though not during pregnancy) and many other herbal teas. Many of their ingredients, such as those of Greek mountain tea (*Sideritis scardica*) or St. John's wort, activate hippocampal neurogenesis and improve ß-amyloid metabolism. On the other hand, you should avoid drinks containing caffeine in the afternoon, as the stimulating effect can last for a very long time for many people. In general, it is important to drink enough water throughout the day to prevent you from having to consume a lot of fluids shortly before bed, causing increased nighttime urination and interrupting sleep.

Physical activity makes us tired and therefore generally increases our need for sleep, and it also increases the nightly release of growth hormone. If we also move in the great outdoors and thus soak up a lot of daylight, we increase the nightly release of melatonin and thus also the quality of sleep.[27]

Of Owls and Larks

But there are also factors that influence sleep that you can hardly get under control even with good sleep hygiene. These include the so-called chronotype. This defines the internal biological clock and performance at different times of the day. As Figure 18 illustrates, the child's chronotype changes as they grow up from early sleepers and early risers (lark type) to late sleepers and late risers (owl type). This means that teenagers are naturally becoming more and more late risers and their chronotype is changing from that of a lark to that of an owl. Only after reaching biological adulthood, from about twenty-five years of age, does one become a lark again with increasing age.

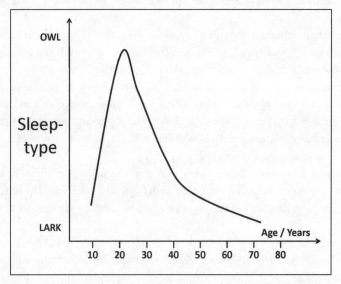

Figure 18: From the lark to the owl—and back. The change in chronotype depending on age.

These chronobiological changes in sleep and circadian rhythms are hereditary and are subject to a variety of gene activities. That's why the graphic only represents average values. Depending on gene variants and individual (inherited) combinatorics, there are sometimes considerable fluctuations in the internal clock.[28] In a larger group or population, this results in a chronotypic normal distribution, as is also known for other human characteristics such as height or weight. In earlier times, it certainly had advantages to have extremely early larks, whose sleeping time is ideally between 9:00 p.m. and 5:00 a.m., and extremely late owls, with bedtimes between 5:00 a.m. and 1:00 p.m., in your clan. These two extremes could provide round-the-clock guarding without requiring those of either chronotype to alter their natural sleep patterns. Perhaps this survival advantage promoted the genetic diversity of sleep regulation and thus the emergence of the two extreme chronotypes and all intermediate stages resulting from genetic combination, which form the majority overall.

This majority has their natural need for sleep, as Figure 19 shows, between 11:00 p.m. and 1:00 a.m. to 7:00 a.m. and 9:00 a.m. respectively. The average natural time to wake up is around eight in the morning, which is exactly when working hours begin or have already begun for most people. Depending on your commute and time for breakfast and hygiene, you may even have to get up two to three hours earlier. However, because people are usually reluctant to go to bed earlier due to their chronotype, because they are not really tired yet, this causes a sleep deficit in the majority of people.[29] The negative health consequences of chronic deviation from the individual sleep rhythm are serious for the individual, but also for society as a whole. For everyone except the extreme larks, it's like waking up in the wrong time zone every day without ever being able to adjust. However, although this has been known for a long time, this problem is still being addressed, perhaps for economic policy reasons.

Conservatism is ignored in terms of health policy. Daylight saving time is a prime example of this, as it has so far been maintained due to business interests, although it exacerbates the problem of getting up too early. The original energy-saving argument that justified its introduction has long been refuted.[30] Instead of doing any good, there

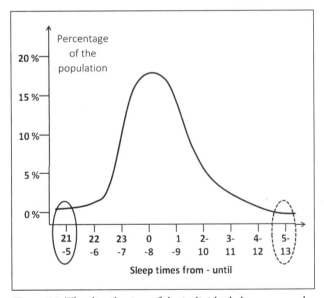

Figure 19: The distribution of the individual chronotype, the times of optimal sleep in society. The extreme early type (sleep from 9:00 p.m. to 5:00 a.m., closed oval) and the extreme late type (sleep from 5:00 a.m. to 1:00 p.m., dashed oval) are part of a broad spectrum of chronotypes.

is a measurable increase in physical and especially mental illnesses[31] and it even leads to an overall death rate of around 3 percent increase for at least a week after the early summer change.[32]

Adults' working hours also influence children's time at school; they should be out of the house when their working parents have to leave. Toddlers are still larks (see Figure 18), but as teenagers they increasingly become owls. This leads to a further and further divergence between the biological clock (the chronotype), which shifts toward sleeping later and thus waking up later, and the "social" clock, which requires children and especially young people to start school unnaturally early. The result is inevitably a significantly shortened night's sleep in young people. As a result, their health and, above all, their mental performance, which is currently in its crucial development, suffers. Studies show that owls are clearly at a disadvantage when it comes to early classwork.[33]

But you can change that, as a large meta-study confirms: After experimental postponements of school start times, the sleep duration of the students lengthened (they did not go to bed later just because they could sleep longer), which improved both school performance and the health of the students.[34] The authors of the meta-study demand—as many others have before, but unfortunately just as futilely[35]—that school times be adapted to human nature and not vice versa, as the schools currently operate, trying to bend nature to the convenience of the school, something that is the impossible: Even with more intensive lighting using daylight lamps, it is only possible to adapt the chronotypical sleep requirement to the social environment to a very limited extent.

According to Juan Manuel Antúnez, psychologist and author of a corresponding study, owls have on average more negative feelings and also have a worse grip on them than larks.[36] They also have more difficulty assessing and processing emotional situations and respectfully asserting their rights when necessary. The reason for this could be that owls are usually constantly forced to neglect their natural sleep rhythm due to work or school. Their ego is therefore chronically more depleted than that of larks. This can also be seen anatomically: Even in young people, the frontal lobe battery is smaller due to later bedtimes and thus an overall shorter nocturnal hippocampal growth phase than in people who have no chronotypical problem with going to bed early.[37] This is why owls also suffer from depression more often than larks, and are more prone to anxiety and mood disorders, substance abuse, personality disorders, insomnia, high blood pressure, asthma, and type 2 diabetes.[38] However, evening people not only tend to suffer from serious physical or mental illnesses more often than morning people, they also die earlier from their consequences.[39] Being woken up during a REM sleep phase (or even waking up due to a nightmare) is unfavorable for ego regeneration and, in the long term, like poison for mental development, worse than a purely quantitative sleep deficit, in which the brain at least tries to partially compensate for the deficiency in the long run through an increased frequency of REM phases.[40] But it is not only frequent interruptions of deep sleep that are problematic for brain regeneration, it also suffers from a shortening of sleep in the morning hours. This particularly disrupts REM sleep, which impairs the recharging and capacity increase of the frontal lobe battery.

In the Race Across America, the nonstop cycling race across the US, most participants suffer from extreme sleep deprivation due to their unnatural racing strategy, sometimes with only one hour of sleep per day over eight to twelve days, and very often they develop paranoia as a result. I suspect that what is behind this is the hippocampus's attempt to upload and dreamily link the new memory contents with previous ones, necessarily while awake. Dream and reality overlap. But even without extreme physical exertion, with severe sleep deprivation, the risk of developing mental problems, from hallucinations to more serious psychoses, increases in general.[41] Vital REM sleep is often not only disturbed by getting up early due to school or work obligations, but also from the outside, so to speak. Some drugs can contribute to this.[42] Ironically, these include the most commonly used sleeping pills. Many people do not solve their sleep problems through better sleep hygiene (which would be the natural way), but instead turn to benzodiazepines, an artificial group of drugs that depress the brain, or active ingredients that manipulate the benzodiazepine receptor in the brain and thus have the same effect trigger. According to a DAK health report from 2017, they are prescribed to one in seven employed insured people because of sleep disorders.[43] Of the older population, up to a third take such medications almost every day.[44] Unfortunately, they hinder REM sleep, which is so crucial for nighttime recovery. The possible effects of long-term use of benzodiazepines include concentration and memory problems as well as the development of depression. Paradoxically, however, this group of substances is often prescribed concomitantly with therapy in the latter.[45] The risk of developing Alzheimer's also increases[46]—most likely due to the chronic disruption of nocturnal hippocampal neurogenesis, which predominantly occurs during REM sleep.

Lack of Sleep Promotes Stereotypical Thinking and Racism

In the acute as well as chronically depleted ego, there is also a feedback blunting, a kind of feedback disorder, which manifests itself in the inability to absorb and implement corrective signals—in other words, to use System II when it could be a mistake to stereotypically act with System I due to an unclear information situation.[47] Fatigue and lack of sleep therefore let us carelessly make mistakes, which repeatedly has life-threatening consequences in clinical emergency measures, disaster management, and military or police operations. Even in socially tense situations, it is problematic when the stereotypical System I dominates because System II does not kick in due to feedback blunting.

If insufficient regeneration of the frontal lobe battery is a chronic problem—as it is for an increasing proportion of the world's population—then feedback blunting or the inability to switch on System II also becomes a significant problem when it comes to, for example, eating a healthy diet. If you're overtired, the chocolate doesn't stay in the cupboard, despite all your good intentions. Chronic fatigue is also becoming a societal problem, as overtired people are less tolerant of others, for example, if they have different religious beliefs.[48] One does not automatically become a xenophobe or racist when suffering from sleep deprivation, but one can suppress prejudices, if they are already part of one's personality, less effectively with a depleted ego.[49] The stronger the ego depletion due to sleep deprivation (and ultimately due to any mechanism that inhibits adult hippocampal neurogenesis), the more narrow-minded (frontally weaker) those affected think and act.[50]

Enough Sleep for Everyone!

Lack of sleep changes our psyche and our mental abilities, but not for the better. In order to minimize the catastrophic consequences for the development and growth of the frontal lobe battery—or even better, to avert them completely—we as a society should ensure that we *all* get enough restful sleep. To achieve this, the start of work and school should be adapted to the evening chronotype. The owls among us would benefit from this, and it would not be a disadvantage for the larks. Starting work and school one or, better still, two hours later would enable a large part of the world's population to lead a physically and, above all, mentally better developed and overall healthier life. Night shifts should be abolished as much as possible, or the times should be adapted as closely as possible to the chronotypes of the respective workers and employees.[51]

In fact, such measures should not constitute an insoluble problem, nor should the urgency of their implementation be seriously questioned. In most professions, it doesn't matter when you start working, as long as you have accomplished your workload at the end of the day. However, if lack of sleep is already preprogrammed and possibly even long-term (as is unfortunately often the case these days), then the goal is wrong. It is important to keep in mind that we work to live, not the other way around. If we are harmed by economic constraints—regardless of what traditions or conservative mindsets they are based on—they must be changed. But the opposite happens. According to psychologist Hans-Günter Weeß, board member of the German Society for Sleep Research and Sleep Medicine (DGSM), home offices and mobile media mean that we will soon be a "24-hour non-stop society."[52] The working world must become more flexible, and parents in particular could be pioneers here. Nevertheless, two-thirds of the parents surveyed even speak out against a later school start, although it should hurt them to tear their children out of deep sleep every morning, and they know about the harmful health effects.[53] I therefore wonder whether the parents answered the surveys with System I or System II. One solution would be to reduce weekly working hours to perhaps thirty and distribute the work itself among more people. This would also reduce unemployment (which would give many people a sense of meaning in life). A later start to work would not automatically mean a later end to work. But maybe that's just a dream. But "responsibility begins in dreams," as the Irish playwright and Nobel Prize winner for literature William Butler Yeats (1865–1939) recognized. I wonder how we as a society can sensibly take responsibility for ourselves and future generations and make sensible and sustainable decisions every day if we rob ourselves of sleep and interrupt our dreams every morning.

CHAPTER 11

EATING SENSIBLY FOR YOUR SENSES

If children eat poorly, they live poorly.
—Henrietta H. Fore, executive director of UNICEF

Culture of Malnutrition

I'm currently devouring a pound of cervelat sausage, with mustard and pickles and half a loaf of white bread. Once I've eaten everything, I'll have managed to make my grandparents happy. My grandpa is already retired, and what my grandma earns as a saleswoman isn't much. But there is no scrimping for her grandson, especially not on sausage. I wash down the salty, fatty taste with sweetened lemonade—there's plenty of that too. And my beloved marzipan potatoes are waiting for me for dessert. My grandfather also likes it sweet, and he and Grandma are thrilled when I shove the first fatty-sweet ball into my mouth after a sumptuous dinner. "The boy sure can eat" sounds like praise coming from my grandfather. I already know that my mother will hear this too. My grandparents see themselves in a competition with my mother, where it's about who makes the food taste better to me. My grandmother pinches my hamster cheeks as if she wants to measure her success, and is satisfied.

My grandparents' idea of what is important for life and survival and what is not was shaped by their war experiences. The famine they experienced at the time had a significant impact on the diet of their "war grandson." The declared goal was rosy chubby cheeks as a sign that I was better off today than they and their son, my father, who was born four years before the outbreak of World War II. In their formative experience, children with some fat on their ribs had a better chance of surviving periods of food shortage. In the postwar years, chubby cheeks became a sign of prosperity—and, paradoxically, also a sign of health. Fortunately, my mother didn't participate in this "eating to get fat" and so my love handles slowly reduced again when I was home, because this was not a matter of course and runs contrary to the cultural development of our increasingly fattening society. In 2020, about half a century later, the German Obesity Society (DAG) lamented on the occasion of World Obesity Day: "Even medical practitioners, healthcare decision-makers and politicians often do not understand or do not want to accept that obesity is a chronic disease."[1]

The prevalence of overweight and obesity has increased worldwide since 1980.[2] The world population is becoming increasingly fat. According to WHO studies, around 2.3 billion people worldwide were overweight in 2016 (almost 31 percent), of which 700 million adults were obese (extremely overweight) and 338 million children and adolescents were overweight or obese.[3] The average weight gain in children and adolescents is even accelerated. According to data from the National Center for Health Statistics, in the United States alone the prevalence of childhood obesity more than doubled and even tripled among adolescents in the three decades around the turn of the millennium, from 1985 to 2015. Overall, 20.6 percent of US adolescents, 18.4 percent of school-age children, and 13.9 percent were already severely overweight at preschool age. This means that obesity in the US has reached epidemic proportions and is affecting the lives of millions of citizens in the long term.[4] But these frightening figures are not only the result and problem of the US fast-food culture—the developments among children and adolescents in Germany and ultimately throughout Europe are comparable.[5] This has serious consequences, especially for young children, because it is precisely the time until school that represents the critical time window for the long-term development of obesity. Obese children are much more likely to become obese adults than those of normal weight.[6]

Sick like Grandpa. Within just two years of observation, the average blood pressure of overweight four- to six-year-olds increased by about 4–5 percent, according to the conclusion of a Spanish study.[7] Hypertension or high blood pressure, previously almost exclusively a problem of the elderly, has become a dangerous consequence of an unhealthy lifestyle in our youngest. In Germany, too, around 8 percent of three- to ten-year-olds suffer from it.[8] This is extremely problematic because high blood pressure leads to blood vessel changes, a disruption in the supply to the brain, and ultimately a reduction in brain mass. Early in life, the frontal lobe and its battery are particularly affected in their development.[9]

Worldwide, more people now die from overeating than from any other disease—even compared to malnutrition.[10] In the US alone, about 400,000 people succumb to cardiovascular disease every year.[11] In addition, there are many more deaths from diet-related cancers, Alzheimer's disease, and vascular dementia, as well as an increased mortality rate from infectious diseases due to a poorly functioning immune system. Overall, many millions of deaths could be avoided through a healthy diet, year after year. Type 2 diabetes alone, as this supposedly age-related diabetes is called, doubles the mortality risk.[12] Germany, with around a thousand new cases daily and a total of almost 10 million type 2 diabetics, sadly holds the leading position in the EU.[13]

It has long been known that cultural information (for example, how to eat) can spread as well as genetic information (for example, in the form of a virus). In the age of the internet, it can do this even faster than before and now influences the lives of almost everyone. Based on genetics, a thought or behavior is called a meme.[14] Like a gene, a meme represents a replicable unit of information that can change randomly (mutation) and spread (selection).

Memes probably influence our lives far more than our genes.[15] In relation to the spread of obesity, for example, it has been found that memetic information about the diet and physical activity of parents has a very strong predictive power for how their children's weight will develop.[16] The eating behavior of parents is also transmitted to their children like a virus.[17] Depending on which meme they "infect" their children's brains with, they have a significant influence on their weight development. But it is the entire culture that has an effect on this, for example in the form of the snacks at the checkouts of the supermarkets or the 12,000 commercials that children in Germany aged six to thirteen see every year on "children's television," which controls their eating behavior.[18] Studies show that the more commercials children see on television or online, the more serious their addiction to junk food becomes.[19] Even short advertisements lead to a measurably increased calorie intake. If the correspondingly advertised snack is already nearby, the effect is even more pronounced, as has already been found with preschool children.[20]

The influence of influencers. According to the "Junkfluencer" report by the consumer protection organization Foodwatch, industry-funded social media stars who successfully promote junk food and unhealthy behaviors influence children and adolescents by exploiting their popularity.[21] "Unfortunately, companies and influencers can earn significantly less with a healthy diet than with unhealthy products," explains Foodwatch. This is "a clear economic disincentive." In this way, companies sabotage "the efforts of many parents to inspire their children to eat healthy." Professor Berthold Koletzko, chairman of the Children's Health Foundation, says, "The subtle, perfidious advertising that families cannot control must be regulated."

According to the results of a controlled study, processed foods increase calorie intake by about 500 kcal per day. In the experiment, this led to a weight gain of around one kilogram after just one week of fast food instead of a balanced diet.[22] The reason for this is the more quickly available sugar and a lack of fiber. This is staggering, because according to the results of an extensive study published in the renowned *New England Journal of Medicine*, every extra pound causes considerable damage when you are overweight. Obesity in childhood alone increases the risk of stroke in later life by five times.[23] Highly processed ready-made meals ultimately shorten the lives of a large part of the population.[24] Yet, despite hundreds of deaths every day in Germany alone, there are neither warnings in the evening news nor serious political measures taken to address this. No lockdown of the fast-food industry is to be expected; not even the advertising of their deadly products is being restricted.

The UGLY. Students in basic economics courses learn that advertising primarily serves to protect the consumer. The logic of this statement is based on the thesis that if there were no advertising, consumers would have fewer purchasing alternatives and this would make it easier for monopolies to emerge, which would then charge arbitrarily high prices for their products. "According to this theory," says a report in *Die Zeit*, "it is not only the companies that are powerful, but also the consumers. They don't have to buy the products advertised. It's their choice."[25] But children neither understand the mechanics of advertising nor do they know what type of diet is harmful to them. Therefore, they are easy and unprotected victims of industrial advertising campaigns. That's why the German food industry alone pumps around three-quarters of a billion euros into advertising for confectionery and chocolate every year, because the food industry has long since identified our children as willing consumers. This is how they directly damage our children's frontal lobe battery. Too much sugar in childhood impairs hippocampal neurogenesis just as dramatically as a severe traumatic experience.[26] An often-lifelong reduced psychological resilience and correspondingly increased stress levels are just some of the many consequences.

In order to protect children from the interests of the confectionery industry, in May 2018, over two thousand doctors (including over 1,300 pediatricians, 222 diabetologists, and 58 medical professors) called on Chancellor Angela Merkel in an open letter to the federal government to finally take political measures to prevent obesity and diabetes and other chronic diseases. It says, among other things: "An unbalanced diet, which is often learned in childhood, is one of the reasons for this worrying development. Children between the ages of 6–11 consume on average about twice as many sweets and sugar-sweetened beverages, but less than half as many fruits and vegetables as recommended by the Research Institute for Child Nutrition."[27] Accordingly, Thomas Fischbach, president of the Professional Association of Pediatricians and Adolescent Physicians, also calls for "an advertising ban on so-called children's foods, which actually do not exist."[28] Because since when do children need different food than adults? From the point of view of Matthias Blüher, president of the German Obesity Society, "the problem can only be solved by taking political action."[29] In his opinion, there should also be a "redirection" that makes harmful food more expensive and food rich in vital substances cheaper. This means every family can afford healthier food.

According to Thomas Danne, chief physician at the Auf der Bult children's and youth hospital in Hanover, even many daycare centers and schools only offer unbalanced meals "for cost reasons, because high-quality food is more expensive."[30] When it comes to the supply of sweets themselves, this also needs to be addressed. In my opinion, the legislature should at least stop the sale of sweets in schools. However, in Germany there is neither a special tax on sugary and fatty drinks or sweets, nor are there any restrictions on advertising aimed at children, although, according to the WHO, such measures have significantly reduced the consumption of unhealthy fattening foods in other countries.[31] Christian Schmidt, the Federal Minister of Food from 2014 to 2018, had repeatedly rejected such measures. In September 2015 he said in the *Tagesspiegel*: "I reject political control of consumption through advertising bans and punitive taxes on supposedly unhealthy foods."[32] Schmidt's successor, Julia Klöckner, also shows no interest in legal restrictions on the industry, despite another alarming report on the obesity pandemic published in *Lancet* in 2019.[33] In it, leading international researchers again demanded that "to get a grip on obesity and malnutrition, the influence of the food industry must be pushed back."

The UGLY. It is understandable that the confectionery industry wants to bring as many of its products as possible to men, women, and especially children; after all, it is always about maximizing sales and profits. Thus, companies worldwide are not only investing millions in marketing to target our children (see also "The influence of influencers"), but also in lobbying to prevent child protection laws, so that their sales continue to rise. How much money is invested in Germany to influence politicians in a pro-industry manner is unclear because the figures do not have to be disclosed in this country (to protect politicians and industry, certainly not to protect consumers). However, they have to in the US, and there they are frighteningly high.[34] This is one reason that measures should be taken to protect our children from the interests of the confectionery industry.

Foodwatch described the Federal Minister of Food as "dangerous to health" because, as usual, it continues to rely only on the voluntary commitment of the confectionery industry.[35] With her soft policy, Klöckner even goes against the demands of the scientific advisory board of her own ministry. She also advocates the introduction of a soda tax, the reduction of the Value-Added Tax (VAT) on fruit and vegetables, the restriction of children's advertising for unhealthy foods and mandatory nutrition labeling with the NutriScore traffic light.[36] These are all measures that have been shown to help children to eat and develop healthier. According to Martin Rücker, managing director of Foodwatch, the minister's failure to act "represents an abuse of office given the massive health consequences for children."[37]

Commercial euthanasia. When the German parliament decided in 2015 that commercial euthanasia (Section 217 of the Criminal Code) would remain prohibited, the satirical news magazine *Der Postillon* reported "numerous fast-food chains—including McDonald's, Burger King, Kentucky Fried Chicken, and Pizza Hut—announced that they would examine the legal situation to determine if they would be forced to close all their German branches."[38] The report continued: "Whether the fast food companies are really affected by the decision is still unclear. The new law literally states that its aim is to 'prevent the development of assisted suicide into a healthcare service.' However, there is still no consensus among lawyers on whether 'healthcare' also includes food in the broadest sense." In February 2020, the Federal Constitutional Court ruled that this could be one of the reasons why, politically, suicide remains largely unpunished. This is considered a positive outcome, as *Der Postillon* suggested that mass closures in the fast-food sector would be disastrous for society "because, according to company reports, hundreds of thousands of Germans eat at McDonald's every day simply because they no longer care about their lives."[39]

Bigger Belly, Smaller Brain

The government acts as an agent of the industry. But perhaps it is only their goal to attract another generation of conservative and market-liberal voters who find it difficult to think system-critically about System II and who are therefore easy to influence. In this (perfidious) case, such a voter is doing everything right if they do not protect our children from the interests and machinations of a malnutrition industry that makes them sick and mentally inflexible. This also worked in ancient Rome. Even then, "bread and circuses" were a common instrument for keeping the common people quiet with simple means. But with the realization that conservative satisfaction could be based on a chronic weakening of the frontal lobe battery, this kind of manipulation takes on a completely different dimension. Perhaps the saying "dumb as bread" already reflects in some way a "felt knowledge" of this connection. After all, malnutrition and obesity cause more than just physical damage. Unfortunately, chronic fat and carbohydrate-rich overeating also lead to a direct impairment of mental health; an increase in body volume is accompanied by a decrease in brain volume.[40] However, this is not to say that someone who has too much belly fat has too little brainpower. You can be mentally above average even if you are overweight, but most likely not in the way you could be if you weren't overweight or even obese. The same applies to physical fitness, which hardly needs any explanation. The average volume of the brain of overweight people is reduced by around 4 percent compared to normal weight people, and even by 8 percent in extremely obese people.[41]

Responsible for the fact that a smaller brain on average is not the cause but the consequence of overweight is, among other things, the visceral fat, the deep abdominal fat lying around the intestines.[42] It has long been known that visceral fat storage plays a crucial role in the development and maintenance of our brain mass, because it is not a passive energy depot, but a highly active hormone gland.[43] By producing and releasing a whole army of hormones and hormone-like substances, visceral fat tissue intervenes in all vital processes of our body. These additional functions are extremely useful from an evolutionary biology perspective, because it is of great advantage for survival if all organs, especially the brain, receive hormonal signals about how much energy is available to the body and even how it should be used. Fat tissue, which has long been mistakenly thought to be silent, loudly determines whether one feels hungry or full, whether the body can initiate energy-consuming regeneration processes or not, whether the immune system has enough energy to attack or should behave rather calmly, and not lastly, whether the organism can afford a larger brain for purely energetic reasons.

What all adipose tissue hormones have in common is that they directly or indirectly control hippocampal neurogenesis. Optimal activation occurs at a medium abdominal fat volume, in the healthy corridor between minimum and maximum. A lack of abdominal fat, but also a chronic excess, overrides this hormonal regulatory system and the hippocampus shrinks. Such a hormonal imbalance promotes inflammation, a neuronal insulin resistance (to which the hippocampus is particularly sensitive, as I will explain later), and also a loss of frontal lobe battery capacity through type 2 diabetes, high blood pressure, and arteriosclerosis. A classic example of this is the failure of the so-called leptin adipometer, our "fat meter" built into the abdomen.

Leptin adipometer and leptin resistance. Leptin is a hormone produced and released by fat cells. Through the leptin level, our belly fat informs the brain how much energy is available or how much fat is stored, because less well-filled fat cells release less leptin than saturated ones. If our fat or energy reserves run low, falling leptin levels make us hungry. A lot of fatty tissue, on the other hand, produces a lot of leptin to signal our brain "loudly" that the stores are full. Appetite decreases, or it should, because if a full memory becomes a permanent condition (as is becoming increasingly common in our modern society), "leptin deafness" develops as a result of the continuous exposure to "sound." The nerve cells then no longer have an "open ear" for leptin, and this is medically referred to as leptin resistance.[44] For our brain it is as if there were no leptin present, even though large amounts of this hormone circulate in the blood. Despite a surplus of stored energy, people are extremely hungry. Leptin resistance is thus harmful, also because many other effects of this hormone are absent, such as the breakdown of no-longer-needed β-amyloid and the stimulation of adult hippocampal neurogenesis.[45] Instead of growing, the frontal lobe battery atrophies: "Being overweight is associated with atrophy [reduction] of the hippocampus," according to the title of a paper by Australian scientists, who also call for political measures against obesity.[46]

A factor that exacerbates developing leptin resistance is chronic sleep deprivation, as this leads to a lower release of leptin and an increased release of the appetite hormone ghrelin.[47] The consequences are an increased feeling of hunger and further weight gain. The sleep deprivation pandemic thus exacerbates the obesity pandemic and turns it into an exacerbated pandemic of frontal lobe weakening. Chronic stress due to a constant lack of time or due to social problems or unresolved trauma also causes many people to overeat—following the idea of "comfort food." If this is the case, one of the best starting points for weight loss is psychotherapeutic intervention.[48] If clinical depression develops as a result of disturbed hippocampal neurogenesis and the resulting reduced psychological resilience, this can also be the cause of an eating disorder, as Figure 20 illustrates, and thus have a self-reinforcing effect.

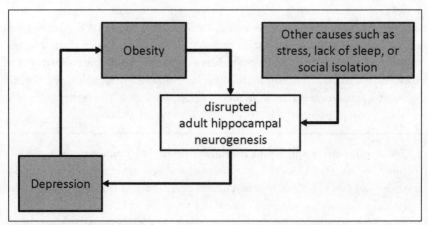

Figure 20: Vicious cycle of obesity caused by impaired hippocampal neurogenesis.

Lack of sleep, depression, and the tendency to impaired lipid metabolism are both consequences and possible causes of chronic overeating. In addition, with an excess of visceral belly fat, the probability of developing dementia in the long term also increases; in fact, it is up to four times higher than in people of normal weight.[49] A twin study from Sweden lasting over forty years revealed, however, that excess weight not only robs us of mental capacity in old age, but weakens it many decades earlier.[50] One reason for the shrinking of the hippocampus with too much visceral fat is—in addition to the lack of growth impulses due to a leptin resistance caused by belly fat—also a deficiency in adiponectin. This is another fat cell hormone that also activates hippocampal neurogenesis, among many other health-promoting mechanisms. However, in obesity, it is released in smaller quantities, which means that its important functions for mental and physical fitness are no longer present.

Another reason for the loss of frontal lobe battery capacity in the case of malnutrition is the already mentioned insulin resistance. Again and again, sharply increased blood sugar levels cause a kind of habituation. The insulin receptor, similar to the leptin receptor, loses its function when leptin is constantly increased: Despite high insulin release, blood sugar remains elevated. Insulin resistance is the main cause of what is known as "adult-onset diabetes." However, with every excess kilogram, the risk of developing this supposedly age-related diabetes increases—even in children. The youngest child was just four years old at diagnosis.[51]

But even with a chronically only slightly elevated blood sugar level, which is not yet clinically classified as type 2 diabetes, both developmental disorders and loss of capacity in the frontal lobe battery are already detectable, including a poorer memory.[52] These findings are of enormous significance. Obviously, repeated intake of sugary foods, even without the development of an abdominal goiter or clinical diabetes, can permanently damage the frontal lobe battery. One reason for this could be the inflammatory properties of sugar. Increased blood sugar levels lead to adhesions on all cell surfaces, similar to a spilled sweet drink. The "sticky" things in our bodies are "advanced glycation end" products, or AGE. In both German and English, AGE coincidentally means "age"— and in fact, AGEs accelerate aging.[53] In principle, AGEs are nothing more than the body's own proteins, fats, and even nucleic acids (which make up our genetic material) that have been chemically altered by sugar (glucose, fructose, corn syrup, etc.) and lose their function due to this sticking. But it gets even worse: The AGEs activate a special receptor for AGE, abbreviated RAGE, which recognizes these adhesions as harmful foreign bodies and activates the immune system to remove them. "Rage" means anger, which describes it quite well—it is the mood of our immune system when RAGE activated by AGEs causes a massive release of inflammatory activators such as interleukin 1, interleukin 6, or tumor necrosis factor alpha, all of which drive the destruction of nerve cells and also massively inhibit hippocampal neurogenesis.[54] Just four days in a row of eating a carbohydrate-rich breakfast such as white bread with jam was enough to observe a significant decline in hippocampal performance compared to a control group that ate a healthier breakfast.[55]

Unfortunately, just avoiding foods that cause blood sugar to rise quickly is not enough to appease the RAGE system. AGEs also enter our body preformed as part of our diet.[56] In particular, meat and dairy products from factory farming are extremely rich in AGEs. The reason for this is the close confinement and fattening of the animals. These conditions lead to constantly elevated blood sugar levels (which promotes the desired rapid weight gain). In a certain way, meat consumption continues the accelerated aging process of the tortured animals, which they themselves no longer experience due to their early death. The already very high AGE concentration is even further increased by the way food is prepared, especially in fast-food production.[57] According to the results of a corresponding study, just one week of junk food as part of the diet is enough to switch off the natural hippocampus-dependent appetite control for several weeks.[58] This closes another fatal vicious circle that makes us increasingly fatter, blocks the growth of the frontal lobe battery, and, by activating the AGE/RAGE system, causes hippocampal neurodegeneration and thus a chronic loss of frontal lobe battery capacity.

The UGLY. Industrial factory farming results in unhealthy, cheap meat and a completely unnatural diet. To make this madness possible, the EU is buying record quantities of soy from the US and South America as a source of protein for animal fattening.[59] Almost 99 percent of global soy production is destined for mass-tormented animals. However, only about 10 percent of the protein building blocks (amino acids) contained in soy end up at the meat counter. The large remainder is excreted by the animals and ends up in the fields as nitrate-rich manure and solid manure—in Germany alone, it was about 204 million tons in 2015.[60] Through our alien lifestyle, we destroy ecologically indispensable rainforests at record speed because the land is needed for soybean cultivation, contaminate our soil and drinking water through over-fertilization of agricultural land, and cause life in lakes and marine regions to die off.[61] In addition, this state-subsidized diet increases the rate of so-called affluence diseases, which range from a significantly increased cancer rate to a veritable mass death due to cardiovascular diseases. According to the results of a comprehensive European study, over half of all fatal heart attacks and strokes are caused by our unhealthy diet.[62] As we have seen, the risk of developing type 2 diabetes also correlates very strongly with the level of meat consumption. However, everything that damages heart and blood vessels also hinders the healthy development and preservation of our mental faculties. The same holds true for the nitrate from liquid manure in drinking water; it impairs early childhood brain development and, in the long term, contributes to

Alzheimer's disease and many other neurodegenerative diseases.[63] But not only do we destroy ourselves and our habitat and inflict unimaginable suffering on animals, we are also responsible for world hunger. The more animal products we eat, the fewer people the Earth can feed. The world's cultivated land is limited and the overexploitation of nature is increasing (to which our hunger for dairy and meat products contributes to a large extent). However, if everyone were to live a vegan lifestyle, no one would have to go hungry and, according to the latest studies, there would even be enough food for up to four billion more people.[64]

Fasting for Fitness

During periods of fasting, blood sugar drops to a healthy level. At the same time, fat stores are broken down in order to provide energy to our organism by means of the released fatty acids. Our brain—based on its relative size—needs around ten times more of this than the rest of the body. However, there is a problem that needs to be solved. The fatty acids stored in adipose tissue and released during fasting are predominantly saturated and, unlike polyunsaturated fatty acids such as aquatic omega-3 fatty acids, cannot cross the blood-brain barrier.

This barrier between the blood vessels and the brain protects its nerve cells from many toxins and disease-causing germs, but also prevents the energy supply from saturated fatty acids. In order to still be able to use these, nature has a trick up its sleeve. They are first broken down in the liver and converted into ketone bodies. These can now effortlessly overcome the blood-brain barrier and thus supply the brain with the energy previously stored in fatty tissue. And we have plenty of it: even one slim person of 70 kilograms, for example, has about seven kilograms of storage fat with a relatively low body fat percentage of only 10 percent. At around nine calories per gram of fat, that's around 63,000 calories. In comparison, our energy supply of sugar, which is stored in the form of glycogen (a type of starch) in the liver and muscles, is extremely limited. Even if their glycogen stores are completely full, an adult only has the daily requirement of around 2,000 calories of energy available, and most of this is in the muscles, which need this energy themselves and do not release any of it to the brain. Thus, the fatty acids stored in our fat reserves, converted into ketone bodies, represent a gigantic energy source for our brain, which would allow even slim individuals to fast for a good month, unlike sugar from glycogen stores.

"Baby fat" also serves to supply the infant's growing brain with energy using ketone bodies. They cover a large part of its energy needs and, as hormones, simultaneously promote neurogenesis.[65] Even in adults, they have this dual function and contribute to the capacity expansion of the frontal lobe battery.[66] The evolutionary advantage of this hormonal effect could be that a fasting organism usually goes in search of food and will therefore have new experiences that a growing hippocampus can better store and process. Ketone bodies not only stimulate neurogenesis, they also activate the regeneration of our nerve cells.[67]

	Ketone Bodies	Sugar
Inflammation	Inhibiting	encouraging
Cell metabolism	rejuvenating	aging
Neurogenesis	activating	blocking
Neurodegeneration	inhibiting	encouraging
Type 2 Diabetes	inhibiting	encouraging
Obesity	reducing	encouraging

Ketone bodies—our best source of energy.

As the table shows, high ketone body levels signal the hippocampus to grow. They also activate immune cells that protect nerve cells.[68] High blood sugar levels, on the other hand, lead to inflammation, which inhibits brain growth. They also make our nerve cells age faster. Ketone bodies improve cell metabolism by activating the breakdown of waste in our cells and thus ensure their rejuvenation. Fasting is therefore healthy because ketone bodies are healthy and protect us from many diseases. But as soon as we consume sugar in the form of bread, pasta, sweets, or sweet drinks, the increase in blood sugar and the subsequent release of insulin blocks the release of fatty acids from the fatty tissue and also stops ketogenesis in the liver. But after about a twelve-hour break from sugar, ketone bodies increasingly replace blood sugar (glucose) as an energy source for our brain. This period roughly corresponds to our nightly sleep, making it the most natural form of fasting. In English, this is also expressed by the word *breakfast*: it literally means "breaking the fast."

The starving hippocampus. An important characteristic of the very early phase of Alzheimer's development is insulin resistance in the hippocampus nerve cells.[69] As a result, they can absorb virtually no glucose, which can be used diagnostically for early detection. If sugar is marked, so that it can be visualized in an imaging procedure, it is evident even before the first clinical signs of Alzheimer's that the hippocampus region is spared.[70] If those affected in this early phase offer their frontal lobe battery only sugar as an energy source, which is the normal case with a modern Western diet, it starves despite high blood sugar levels. This significantly speeds up the disease process. Only ketone bodies can continue to supply the hippocampus with energy—and, due to their hormonal properties, can even protect us from Alzheimer's.[71]

Despite all the benefits of ketone bodies for our brain metabolism, most people in our modern society produce next to none. The reason for this is the intake of sugary foods in too-short intervals (breakfast, lunch, dinner, and snacks in between) and a nightly eating break that is usually far less than twelve hours. It would be quite easy to change this in a brain-friendly way. Nighttime fasting can be extended by eating your last meal a few hours before you go to bed and possibly skipping breakfast or postponing it until later in the morning. Even the former fisher and gatherer didn't go to the refrigerator first and eat a sausage or jam on toast for breakfast, but rather had to go out into nature to find his food. Our heritage is still well adapted to this today. Even without a full stomach, we are highly productive thanks to ketone body production.

If you would like to use your fat storage in the form of ketone bodies as a hormonally active energy reservoir in the future, you should know that the ketone body metabolism needs about one to two weeks to get fully going again. The nightly fasting breaks gradually become longer and longer without feeling hungry. Once you get used to it, you are even mentally fitter at work without your morning sugar intake, because ketone bodies are a more efficient source of energy for the brain than glucose, which means that they provide more energy with the same oxygen consumption (this is especially important when the blood supply to the brain is impaired). On the other hand, when you eat a large breakfast, rising blood sugar levels cause insulin to be released, which excessively reduces the only energy source available, causing fatigue and further appetite. Even in primary school children, quickly available carbohydrates fed as breakfast, for example from fruit juices, have been shown to cause poorer performance in reading and arithmetic.[72] Meta-studies show this: The lower the blood sugar rise after breakfast, the better children and adolescents perform at school.[73] But regardless of the time of day, fruit juices (like all sugary drinks) and rapidly available carbohydrates (like overly sweet snacks) are generally unhealthy for the development of our children's mental fitness.[74]

If you limit the daily time interval of food intake to about ten hours or fast for about fourteen hours overnight, ketogenesis has been shown to reduce body weight and blood pressure and improve cholesterol and fat metabolism.[75] This protects against obesity, type 2 diabetes, and arteriosclerosis, and since ketone bodies also activate neurogenesis, it also protects against depression and Alzheimer's. But even during the day, when we are no longer fasting, ketogenesis can be maintained by, for example, using virgin coconut oil as a substitute for butter and oil for frying and baking. Over half of coconut oil consists of medium-chain fatty acids. These are water-soluble (in contrast to long-chain fatty acids, they get from vegetable oils or sausages and cheese) and after digestion via a special bloodstream (the portal vein, which connects the intestine with the liver) directly into the liver, where they are immediately and very efficiently converted into ketone bodies. Since coconuts were available as an energy source early in human evolution, it could be that the ketogenic fatty acids from food—first from breast milk and then from coconuts—helped spur the evolution of ever-larger human brains. Presumably, according to palm researcher Hugh Harries, the development of *Homo sapiens* is more closely linked to this palm fruit than previously assumed.[76] This is also supported by the fact that our metabolism, and here in particular the hippocampus, is better adjusted to using the metabolic products of medium-chain fats, such as ketone bodies, and at the same time—in contrast to the rest of the brain—the utilization of sugar via its insulin receptor is tightly regulated or limited.

From breast milk to coconut oil. In nature, medium-chain fatty acids occur in large quantities almost exclusively in certain palm fruits and in the milk of some mammal species. Human breast milk has a particularly high concentration of medium-chain fatty acids, at about 10 percent.[77] Thus, fatty acids that are specifically produced by the mammary gland during breastfeeding promote infant brain development.[78] From an evolutionary biology perspective, this was crucial for the enormous brain development of humans.[79] Coconut oil extends the brain growth and regeneration effect to the rest of life in a completely natural way and even protects against Alzheimer's.[80]

Another advantage of coconut oil is that, in contrast to dairy products from ruminants, it does not contain any trans fats and does not form any when heated because it is heat-stable. In contrast, trans fatty acids are formed during frying and deep-frying with heat-labile oils from sunflower seeds, corn germs, or canola, as these consist mainly of polyunsaturated fatty acids.[81] In addition to trans fatty acids, even more harmful products are formed when these oils are heated. Such products are used in experimental research to trigger brain diseases such as Alzheimer's in animals. For example, many fried snacks contain acrylamide—a neurotoxin that also inhibits hippocampus growth.[82] Also, 4-Hydroxynonenal (HNE), which is highly toxic to our brain, and many related toxins are formed when polyunsaturated fatty acids are heated.[83] Therefore, do not use vegetable oils that are rich in unsaturated fatty acids when frying.

The UGLY. The many health benefits of coconut oil are a thorn in the side of US agricultural industries. They do not generate their sales from products made from coconut palms, but from butterfat from foreign milk and oils from corn germ, canola, and sunflower. Perhaps this is why there are always doctors and scientists who, despite this, declare coconut oil to be "pure poison,"[84] even though the data suggests otherwise.[85] The largest clinical study to date, involving approximately 135,000 patients with stable ischemic heart disease (a disorder of blood flow to the heart without acute symptoms), even showed that saturated fatty acids, such as those in coconut oil, protect against heart attack and stroke. This contradicts the now completely outdated belief that saturated fatty acids are unhealthy. Nevertheless, this misjudgment is still propagated and remains firmly in the minds of many doctors and nutritionists. According to the authors of the study mentioned above, "the global nutritional guidelines should be reconsidered."[86] A meta-analysis also came to the same conclusion, but this is also largely ignored.[87]

The WHO estimates that the consumption of trans fats alone kills over half a million people worldwide every year, solely due to the resulting cardiovascular diseases.[88] However, these alarming figures do not lead to laws that protect consumers from this danger. According to the German Nutrition Society (DGE), there is still "no legal requirement to declare the amount of trans fats contained in food products."[89] On the contrary, according to an EU regulation from 2006, "nutritional and health claims that are intended to inform the consumer about low trans fat content are even prohibited."[90] Perhaps this ban is intended to ensure that companies that include trans fats in their products do not have a competitive advantage.

Developmental Deficiency due to Malnutrition

At least 165 million children worldwide suffer from malnutrition, particularly in South Asia and sub-Saharan Africa. According to a study published in *The Lancet*, around 3.1 million people die from this every year.[91] This tragedy is further compounded by many millions of children who, while not starving, can never reach their intellectual potential due to malnutrition and thus have little chance of a better future. Malnutrition in mothers alone contributes to stunting fetal growth, which further increases the risk of premature birth and even stillbirth. However, even if an infant survives birth, its development and viability are at risk because its malnourished mother cannot breastfeed it sufficiently.

But it is not just a caloric deficiency that causes physical and mental developmental disorders and increased child mortality. The fatal consequences of a chronic deficiency of essential micronutrients are just as serious and even more common. For example, around 250 million preschool children worldwide suffer from vitamin A deficiency.[92] Of these, up to half a million go blind each year, of whom about half die over the next twelve months, often from complications caused by the deficiency of essential micronutrients.

Many other deficiencies in trace elements and vitamins, which we will discuss later, are the cause of such infinite suffering and at the same time would be so easy to remedy that it is completely incomprehensible why it has not already happened. But the completely unnecessary deaths continue year after year, and those responsible are easy to identify. As already explained above, the lifestyle and affluent diet of the rich industrialized nations not only make their own population sick, they also ensure fatal undernourishment and malnutrition among all those who are not so privileged and should not be so because of capitalist market principles. If the people in the developing countries of the Global South were financially independent and therefore masters of their own fate, they would not have to and would not allow the exploitation of their labor or the overexploitation of their resources.[93]

As if all this were not bad enough, the true extent of contempt for humanity becomes apparent in times of crisis, such as during the COVID pandemic. Despite the global appearance of the COVID-19 virus, the preventive actions of rich nations to contain the pandemic are only of a national nature. Thus, the immense flow of funds released flowed only as economic aid to companies affected by the lockdowns, while at the same time, according to investigations by the United Nations Children's Fund (UNICEF), funds for development aid were drastically cut.[94] Therefore, there was already a reduction of about 30 percent in funding when COVID-19 broke out, which fell to 75–100 percent during the first lockdown, resulting in humanitarian crises in the vulnerable countries.[95] Thus, globally speaking, exactly the opposite of what was officially proclaimed as the goal happened, namely to save as many people as possible from death as a result of the pandemic. But one should also include those who died (and will continue to die) of hunger and malnutrition due to the failure to provide assistance as victims of the pandemic (or the measures taken). Thus, the International Food Policy Research Institute (IFPRI), which develops strategic solutions to reduce poverty in developing countries, assumes that due to the pandemic measures of the rich nations, an additional 140 million people worldwide had to live on less than 1.90 US dollars a day in 2020 and thus in extreme poverty.[96] According to the United Nations World Food Programme, the number of people threatened with starvation in 2020 was more than 270 million people.[97] This is a doubling compared to the number before the pandemic.[98] The drying up of the already far too low flow of money into developing countries thus became a death sentence for many millions more children and adults. All of this is happening while, paradoxically, people in countries with high per capita income are actually gaining body weight on average as a result of the COVID measures, which increases their risk of a fatal infection.

As the graphic illustrates, rich countries' national measures against the virus disproportionately increase the risk of death from malnutrition in poor countries, and paradoxically this death usually occurs independently of infection. Under the premise of saving people from COVID, more people worldwide die as a result of the measures taken in Western countries than from the virus disease itself—except that these humanitarian catastrophes were not heard of in the main news or in the general media. This lack of empathy combined with a high level of collective narcissism could be the result of a chronic weakening of the frontal lobe battery. How else can such antisocial behavior be explained?

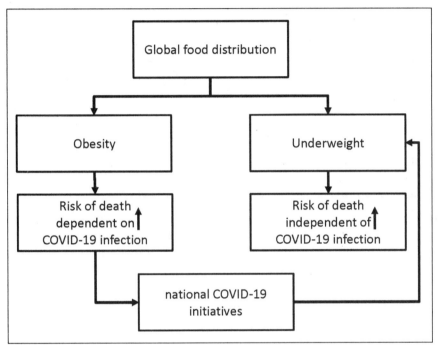

Figure 21: COVID-19 and the deadly consequences of malnutrition and national anti-COVID measures.

Cow Milk Mania

Drinking milk from another species is not appropriate, nor is it appropriate to consume milk from one's own species after infancy. To ensure that this would not happen, nature "punished" anyone who tried it with diarrhea and flatulence for millions of years. The reason for the intolerance of milk (including milk from one's own species) is the loss of the ability to split the milk sugar (lactose) contained in milk into glucose and galactose, which occurs toward the end of infancy. Lactose enters the large intestine undigested and becomes food for intestinal bacteria. When digested, these release carbon dioxide, which explains the unpleasant symptoms. It was not until about three thousand years ago that an accidental gene mutation prevented this natural switching off of the lactose-splitting machinery.[99] Repeated crop failures and famine ensured selection of these mutants. Those who could drink the energy-rich milk of their domesticated ruminants (cows, goats, sheep) as emergency food without getting flatulent had a greater chance of survival. The question of whether this drink, which in itself was unnatural, was healthy did not arise at the time, because the primary aim was to survive. But today we have a choice.

This is not least about the welfare of the animals. By avoiding meat and dairy products, we could save the over four million dairy cows that serve the dairy industry in Germany alone from endless suffering. In the stables, which are usually far too narrow, they have to give birth to a calf every year after artificial insemination so that their milk continues to flow. But they are separated from the natural recipients of their mother's milk, their calves, after birth. The up to 30 liters of glandular secretion a high-performance cow produces every day is not intended for the calf, but for human consumption.

The treatment or prevention of chronic inflammation of the oversized udder requires the massive use of antibiotics.[100] As a result, multiresistant germs develop that also endanger human life, since infections with these pathogens can no longer be treated with medication and therefore often end fatally.[101]

Multiresistant bacteria, but no lockdown on factory farming. Currently, at least twenty thousand Americans and twenty-five thousand Europeans die every year from multiresistant germs, most of which develop through the mass use of antibiotics in factory farming.[102] Worldwide, seven hundred thousand people do.[103] Almost nothing is being done to combat this microbial threat, which could be averted by moving away from factory farming. The use of antibiotics in animals will still be permitted—not only for disease prevention or treatment of the animals, but above all to accelerate their fattening.[104]

Figure 22: Torture over milk. The image of the happy model cow in the alpine meadow is deceptive when it comes to cost-effective mass production of milk.

This madness is accepted in Germany because of an annual turnover of 26 billion euros—and because of the belief in something that the dairy industry has been suggesting to us from an early age with massive advertising expenditure: Milk is healthy. But the data from industry-independent scientists speak against this. For example, carriers of the lactose intolerance mutation, a mutation that allows one to continue drinking milk without the negative effects of lactose intolerance, very often suffer from a different reaction to cow's milk, an allergic reaction, because the immune system recognizes the milk as foreign.[105] However, even more problematic are the chronic effects of milk consumption, which unfortunately are only perceived when it is too late. For example, the research group led by Nobel laureate Harald zur Hausen at the German Cancer Research Center (German: *Deutschen Krebsforschungszentrum*, DKFZ) discovered that cow's milk contains so-called "bovine milk and meat factors" ("meat" because they are also found in beef).[106] BMMFs are novel infectious agents that are responsible for the known increased rate of colorectal and breast cancer in people who drink milk. The DKFZ therefore advises against feeding infants with cow's milk products unless you manage—in an industry-friendly manner—to "humanize" these foreign specie products by enriching them in the future with protective factors such as those found in human breast milk. Alternatively, according to speculation expressed entirely in the interests of the pharmaceutical and food industry, newborns could be given a vaccine against BMMFs, which would allow their parents to continue feeding them nonspecific milk with a reduced risk of cancer.[107] But these are all future scenarios that show how absurdly we live our lives and what we are willing to do just so we don't have to stop with such madness.

The protective factors in breast milk not only block the dangerous BMMFs and thus protect against cancer, they also inhibit the growth of fungi that are dangerous to humans and thus protect against the skin disease diaper rash.[108] Due to the BMMFs present in it and the resulting lack of protective factors, cow's milk even promotes this fungal disease, which is common in infants.[109] Due to skin rashes and massive itching, feeding milk from another species thus causes sleep disturbances in infants and their parents.[110] Since nighttime sleep is important for the development and regeneration of the frontal lobe battery, breastfeeding also promotes species-appropriate brain development and at the same time a healthy or at least undisturbed parent-child relationship.

If, despite all this, your doctor advises you that you and your children should drink cow's milk, arguing that it is good for your bones, then he is either not up to date or a victim of dairy industry propaganda. The reasoning behind this topic is that cow's milk contains a lot of calcium, which bones actually need. However, until three thousand years ago, before the beginning of the cow's milk era, humans were able to obtain all the minerals needed for bone growth from natural sources. This is also possible for all other living creatures that have a skeleton without drinking foreign milk. The majority of people who are lactose intolerant and therefore cannot drink foreign milk still manage to do this today, and generally better than cow's milk drinkers. Not only is drinking cow's milk completely unnecessary for stable bones, it actually has the opposite effect: It softens the bones. One reason for this is the fact that cow's milk proteins are acidifying, which is why our organism has to release calcium phosphate from the bone substance to compensate for this. The phosphate serves to buffer the acid, but the excess calcium is excreted. As a result, bone substance is actually broken down. In addition to the acidification caused by the milk proteins, the galactose released during the breakdown of milk sugar is also thought to be responsible for chronic bone loss. This "mucus sugar" has a very strong tendency to form AGEs, which cause an inflammatory reaction throughout the body, which inhibits not only the formation of new nerve cells but also bone formation. Milk does not reduce, but rather increases, the risk of bone fractures due to osteoporosis (bone loss), as a Swedish research group found.[111] An African study also came to the same conclusion: In lactose-tolerant people who, due to the above-mentioned mutation, were able to break down lactose throughout their lives and therefore generally drank foreign milk throughout their lives, the risk of bone fractures in old age was eighty times higher than in lactose-*intolerant* people.[112]

Galactose also accelerates the aging and breakdown process via the aforementioned AGE mechanism, even if a person only consumes small amounts of milk. This applies not only to the bones, but also to the brain. As has been shown in animal models, galactose inhibits adult hippocampus neurogenesis and even causes neurodegeneration, resulting in chronic memory loss.[113] This property of galactose is also used in animal models to experimentally cause Alzheimer's disease.[114] In mice, a sufficient amount is 0.1 grams per kilogram of body weight daily. In humans, this would correspond to a daily dose of around six to ten grams of galactose or 0.3 to 0.6 liters of milk. It has now been shown that these results of animal experiments can be transferred to humans. Thus, the risk of Alzheimer's increases proportionally to milk consumption.[115] Since it can be assumed that this brain-damaging mechanism affects all brain regions that are in development, including the frontal lobe, this could be one reason why children who are naturally breastfed develop an IQ that is up to five points higher.[116]

But it's not just the unnatural intake of galactose via cow's milk after breastfeeding that is problematic for mental health. The high trans fatty acid content of up to 10 percent of the fat content of foreign dairy products is also harmful to the body and mind.[117] The largest part of the animal trans fatty acids is the vaccenic acid (from the Latin *vacca*, the cow). This trans fatty acid, described as "natural" by the dairy industry, is produced by bacteria in the stomach of ruminants. Vaccenic acid is natural for the cow, but for us and our cholesterol metabolism it is just as detrimental as industrially produced, "unnatural" trans fatty acids, which are created when vegetable oils are hardened and find their way into many finished products.[118] To minimize the risk of a heart attack, stroke, or vascular dementia, it is worth avoiding dairy products for this reason alone.[119]

Animal trans fats are just as dangerous to health as industrial ones. But it's a lot of money and a huge industry. That's why the dairy industry always tries desperately to either prove the opposite or at least sow doubts about it. If a general ban on trans fats in food, as the WHO has long been calling for, would also affect fatty dairy products (or fatty meat products from ruminants), this would be the end of this industry. For this reason, lobbyists from the influential dairy industry ensure that dairy products are always excluded from all regulations on trans fatty acids in food.

Cow's milk also increases the risk of developing type 2 diabetes, not only because of its trans fats, but also because the specific amino acid composition of its proteins has a negative effect on the insulin-producing cells of the human pancreas, as do bovine-specific nucleic acids (bovine microRNA), which are abundant in cow's milk.[120] These also disrupt the natural regulation of our sugar metabolism. Furthermore, cow's milk promotes type 1 diabetes.[121] In this autoimmune disease, which occurs in childhood, the pancreas's insulin-producing cells are destroyed. The amount of cow's milk consumed per day is proportional to the risk of disease.[122] Conversely, the more breast milk an infant has received, the lower it is. The same applies, by the way, to multiple sclerosis, also an autoimmune disease, which can be triggered by the consumption of milk from other species.[123] Interestingly, aquatic omega-3 fatty acids reduce the risk of an autoimmune disease like type 1 diabetes.[124] These are found in human milk, but not in animal milk.

Dairy products are rich in protein to help calves gain muscle mass quickly. But cow's milk proteins are harmful to human brain growth, in particular inhibiting its maturation, studies have shown.[125] So if you don't feed your child like a calf and if you don't want to expose yourself to health risks, stop consuming milk, including dairy products, in your family. And if you don't have children yet, but want to have them someday, then you should consider that cow's milk even inhibits human fertility and thus reduces your chances of ever having children.[126] The reason for this is steroid hormones and many growth factors specifically evolved for the calf in dairy products. Another reason for fertility problems is the contamination of dairy products with pesticides and hormonal chemicals, which have a negative impact on the development of eggs and sperm.[127] All of this also increases the risk of cancer, in women in the uterus and in men in the prostate.[128]

The UGLY. In November 2020, the EU Commission's Agriculture Committee decided that the term "milk alternative" for products such as spelt, oat, almond, nut, coconut, soy, or rice milk is prohibited or remains prohibited. Even the designation "cheese-flavored" for corresponding vegan preparations is prohibited, even if they fulfill the same function as the "real" dairy products.[129] The fear of the dairy lobby is probably too great that otherwise too many people might get the idea to eat in a more animal- and environmentally friendly way, not to mention healthier. The fact that the political elite is again putting the economic interests of a large industry ahead of those of an (even) smaller industry, but above all the common good, is hard to understand in COVID times, when even basic rights are being suspended for the sake of health.

Species-Appropriate Nutrition in the Modern Age

Malnutrition doesn't only prevent children in poor countries from fully developing their intellectual potential. Brain development and frontal lobe battery growth are also disturbed where people eat according to the Western fast-food standard.[130] The cause is an excess of unhealthy fats and unhealthy sugar or toxins produced during industrial preparation. However, a deficit of essential nutrients is just as big a problem.

To find out what our brains are lacking today, it helps to look back to the time of fishermen and gatherers, when the human brain experienced its greatest and final growth spurt. As we have already discussed, many factors contributed to this, not least the pesco-vegetarian or pescatarian diet. This was, among other things, very rich in brain building blocks in the form of aquatic omega-3 fatty acids. Even today, this diet goes well with our basic genetic makeup and gives us the longest life expectancy. As the results of a large study show, pescatarians are, on average, significantly healthier than people who often eat meat and dairy products instead of fish and seafood.[131] Pescatarians also live longer than vegans who completely avoid fish and seafood. One reason for this could be that vegans also have the lowest omega-3 index of all dietary groups, with an average of less than 4 percent[132]—a clear confirmation of the finding that we cannot efficiently convert the plant-based precursor of aquatic omega-3 fatty acids, such as those found in flaxseed or walnut oil.

Fish and seafood provided our ancestors not only with the aquatic omega-3 fatty acids, which are essential for brain maturation and brain health, but also many other vital substances, to the availability of which their genetic program adapted. That's why they are still indispensable today for optimal brain development and the lifelong preservation of our mental strength. These include many minerals such as iodine, selenium, copper or lithium, and many vitamins, such as vitamins A, B_1 to B_7, B12, and D. Interestingly, the amounts required according to the law of minimums are all around 150 to 300 in a daily ration of fatty sea fish grams included. Fiber and some other vitamins that are not found in sufficient quantities in fish and seafood, such as vitamins C, K, or B_9, folic acid, were obtained by our ancestors through their plant-based diet. Even the supply of many minerals essential for brain development and the lifelong capacity expansion of the frontal lobe battery, such as magnesium, is ensured by a vegan whole food diet.[133] Thus, the combination of fish-rich and plant-based foods provides everything for good brain development and is also further evidence of the pescatarian origin of human brainpower. As a result, any individual or combined deficiency in any of these essential nutritional components can become a huge problem for the maturation and maintenance of our ability to think. This also applies to our mental resilience, which is a major reason why expectant mothers who suffer from a chronic lack of aquatic omega-3 fatty acids often suffer from prenatal depression and later from postpartum or postnatal depression.[134] Their risk is increased due to the fact that in the last trimester of pregnancy the need for omega-3 fatty acids for the number of brain building blocks required for hippocampal neurogenesis increases massively in the maturing child (see Figure 11 in Chapter 7). The fetus's needs compete with those of the expectant mother and cause a growth disorder of her frontal lobe battery and thus reduced psychological resilience. Thus, the depression rate during pregnancy and breastfeeding (aquatic omega-3 fatty acids are also passed on to the infant via breast milk) is only about 0.5 percent in those countries where fish and seafood are traditionally consumed in large quantities.[135] In countries where fish is rarely on the menu, on the other hand, the risk of depression increases to over 24 percent during this time, which corresponds to an approximately fifty-fold increase in risk. This clear relationship between a low-fish diet and a massively increased likelihood of suffering from depression was confirmed by another study.[136] It was shown that the severity of depression is directly related to the extent of the deficiency in aquatic omega-3 fatty acids.

The maternal placenta attempts to adjust the fetus's omega-3 index to an optimally high value of 10 to 11 percent for its brain development.[137] An existing deficiency of aquatic omega-3 fatty acids in the expectant mother therefore also causes an undersupply in the fetus. The consequences become apparent at birth: Not only does it occur too early more often, the birth weight is also significantly lower, even if the child is born on the expected date.[138] Developmental disorders can also be observed. For example, in the case of omega-3 deficiency, mental and psychomotor developmental delays can be detected as early as six months of age.[139] According to the results of a long-term study, children's intelligence at the age of four years is lower in children of mothers who were taking an oil during pregnancy that had aquatic omega-3 fatty acids.[140] Based on these results, I can only advise every expectant mother to consume sufficient amounts of an oil rich in aquatic omega-3 fatty acids, ideally algae oil (more on that later).

Competition for the maternal omega-3 store (during pregnancy and breastfeeding) leads to a growth disturbance not only of the maternal frontal lobe battery but also of the child's—with the consequence of reduced psychological resilience and the risk of depression in both.[141] The increased stress perception then also affects the relationship and long-term bond with each other, as illustrated in Figure 23. If the cause is not resolved, the problem becomes self-reinforcing and can escalate. The development of ADHD due to a deficiency-related delayed brain maturation is also becoming more likely.[142]

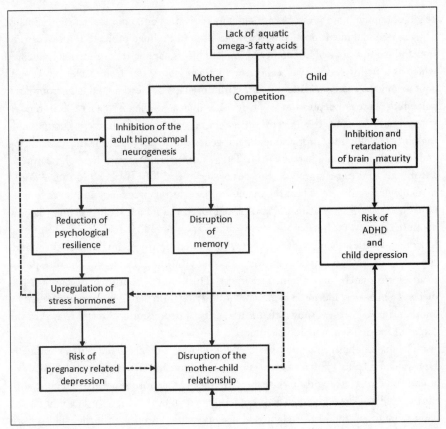

Figure 23: Competition for aquatic omega-3 fatty acids.

Aquatic omega-3 fatty acids, as previously shown in frontal lobe development, are not only important in absolute amounts. Your diet should also have a balanced ratio of omega-6 fatty acids. These also provide important brain building blocks. However, the messenger substances (cytokines) created from omega-6 fatty acids, in contrast to those from omega-3 fatty acids, do not have an anti-inflammatory effect, but rather a proinflammatory effect. That's why omega-6 fatty acids, taken in both absolute and relative excess, inhibit brain growth. For example, it was found that children from countries such as Japan, Korea, and Singapore with an almost balanced omega-3/6 quotient in breast milk performed best in the PISA study.[143] In general, high omega-3 values are associated with good academic performance. However, high omega-6 levels, which are typical of the modern Western diet with a lot of meat products as well as omega-6-rich vegetable oils, such as sunflower or corn oil, significantly worsened the chances of doing well in school.

Toxic food universe. According to the law of maximum, a diet with too many omega-6 fatty acids is harmful to health and mental performance and requires a reduction in omega-6-rich foods. "Individually, a chocolate, a hamburger, or a box of breakfast cereal is not poisonous," reads an interview in the *Neuen Zürcher Zeitung (NZZ)* with Paul Clayton, a nutrition expert and former advisor to the British government, "but in its entirety, our food presents itself with a highly pathogenic [disease-causing] ratio of omega-6 to omega-3 fatty acids and with insufficient fiber, vitamins, and minerals."[144] According to Clayton, this food "causes a climate of chronic inflammation in our bodies, which manifests itself in the form of obesity, heart disease, and diabetes, but also in an increase in psychological ailments and neurodevelopmental disorders." When asked by *NZZ* journalist Theres Lüthi whether the food industry would eventually produce healthier food, Clayton replied: "Only if it's forced to."

Interestingly, when predicting children's future academic performance, the omega-3/6 quotient in breast milk was a more accurate and a clearer forecasting factor for school performance than, for example, country-specific per capita income, government spending per child on school education, or any other measure that has been used and examined to date. This means that if we care about our children's mental performance, it would be better to primarily ensure an adequate supply of aquatic omega-3 fatty acids than to invest exclusively in the education system. Instead of providing the best computers, it would be advisable to improve children's brain health. A primarily pescatarian diet would be ideal for fully developing and maintaining the intellectual potential that lies dormant in your genes throughout your life. But here humanity is faced with a huge problem. As the world's population grows, fishing quotas decrease year after year, meaning there is no longer enough fish for everyone. The average value of the omega-3 index is only about 5 percent worldwide (trend decreasing), and thus provides reason enough for the continuous capacity reduction of the frontal lobe battery in an increasing part of the world population.

Humanity Is Running Out of Fish

To achieve an omega-3 index of at least 8 percent, an adult needs about two grams of aquatic omega-3 fatty acids daily. Ten to 11 percent, as the mother's placenta tries to provide for the supply during the brain maturation of the growing child, would be even better. For a fatty sea fish, this would require around 150 to 300 grams per day. Hardly anyone eats that much fish, and there are three reasons for that. First, most people do not know that they suffer from a huge deficit in this essential brain building material.[145] Second, fish consumption to this extent is no longer recommended because the introduction of chemicals, pesticides, plastic, and heavy metals on an industrial scale has polluted the world's oceans and thus also the marine food chain. And third, over-fishing and environmental destruction contribute to the fact that there is fundamentally no longer enough fish to adequately supply the world's population with aquatic omega-3 fatty acids. Even with fair distribution, given current global production, the average omega-3 index in the world population would remain below 5 percent, as Figure 24 illustrates.

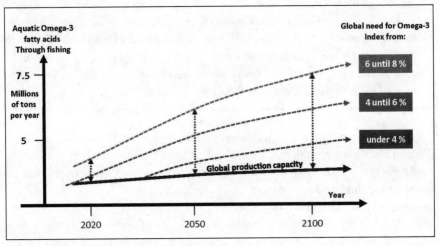

Figure 24: Global Deficiency of Aquatic Omega-3.

By about 2030, the global average omega-3 index will fall below 4 percent due to stagnating production and simultaneous population growth. The up-and-coming aquaculture industry cannot solve the global shortage of aquatic omega-3 fatty acids either, because fish—like us—cannot produce these fatty acids themselves, but must absorb them from plankton. Humanity needs a new source—otherwise, the worldwide loss of frontal lobe capacity will continue to advance unstoppably for this reason alone.

The Algae Oil Revolution

In order to solve the huge problem of the global undersupply of aquatic omega-3 fatty acids, we have to take a new approach. This is possible because, since the beginning of life, the entire world production of aquatic omega-3 fatty acids has taken place almost exclusively in single-celled plant plankton, the microalgae. These microorganisms can be grown in large containers without pollutants, similar to single-cell yeast in brewing beer. The aquatic omega-3 fatty acids obtained from microalgae in the form of algae oil is equivalent in its effect to those from krill or fish.[146] However, algae oil is vegan, free of pollutants, and can be produced on an industrial scale without limitations, so there should no longer be a shortage of aquatic omega-3 fatty acids.[147]

Since aquatic omega-3 fatty acids are fundamental for the development and maintenance of our mental fitness, I advocate giving algae oil the status of a basic food. This means that every person would have the right to a basic supply of aquatic omega-3 fatty acids, just as they have the right to clean air or clean drinking water. For these reasons, I recommend government-subsidized programs that can make algae oil available to everyone at low cost. This is the only way to improve the development opportunities of all children. In this way, one would also counteract the later capacity loss of the frontal lobe battery—with all the negative consequences for the individual and society. Ultimately, this would promote the System II–dependent thinking and action that is so urgently needed in today's world. Cardiovascular diseases as well as vascular (stroke) and hippocampal dementia (Alzheimer's) and even cancer would very likely no longer be the leading causes of death if there was a global basic supply of algae oil.[148]

The UGLY. Experts know that an omega-3 index of 10 to 11 is optimal for mental development (any value below this would already represent a restriction).[149] Nevertheless, official recommendations are far lower. Instead of the two to three grams of aquatic omega-3 fatty acids (DHA and EPA) daily that would actually be necessary, only just under half a gram is recommended for pregnant and breast-feeding women. The German Nutrition Society speaks of 0.2 grams of DHA that should be consumed through fish, which roughly corresponds to this small total amount.[150] The reason for this recommendation, which is far too low, is the high level of pollutants in the fish. The healthy alternative would be contaminant-free algae oil, which contains the same omega-3 fatty acids that have accumulated in fish. But here too, the European Community limits the amount for pregnant and breastfeeding women to 0.45 grams per day by law.[151] This makes no sense at all, especially since fish oil is permitted for pregnant and breastfeeding women with up to three grams per day—unless they want to protect the fish from the algae oil industry. But at what price?

Algae Oil as a Complete Fishing Replacement

With algae oil as a new staple food, we would take a decisive step toward meeting the basic needs of our "pescatarian" brain. But further improvements would have to be made, because after all, fish and seafood contain many essential nutritional components that have also contributed to our brain being able to achieve such high levels of performance. The main reason for this is that sea fish and seafood in particular are the main food source for trace elements such as zinc, iodine, lithium, and selenium, as these minerals have been washed out of the soil by rainwater and washed into the sea in the course of the earth's history. This is why they usually only occur in trace amounts on agricultural land, which explains why they are often only found in plant-based foods in low concentrations. In order to supply ourselves with sufficient quantities of these essential nutritional components in times of global fish shortage, we have to take new approaches. Fortunately, as with aquatic omega-3 fatty acids, there are healthy alternatives for all other essential vital substances that humans have previously obtained primarily from fish and seafood. Below I have stated the daily requirements for adults; you will find the quantities for all age groups in the respective bibliographical notes.

Zinc is an indispensable trace element for maintaining the frontal lobe battery capacity.[152] However, it is also fundamentally vital. A zinc deficiency is responsible for high child mortality in some regions of the world. In 2004 alone, a zinc deficiency cost the lives of around half a million people.[153] Seafood such as oysters contain the highest amounts, but it is no longer possible to consume them frequently enough. Although legumes and wheat bran still have relatively high concentrations, zinc from plant sources is less absorbed.[154] Therefore, it is not easy to meet the daily requirement of around 10–15 milligrams of zinc on a purely vegan diet.[155] I therefore recommend that everyone get their zinc levels checked and possibly take a nutritional supplement. This is so inexpensive that one wonders how humanity can allow so many children around the world to become ill and die from something as banal as zinc deficiency.

Iodine is an essential component of thyroid hormones, the functioning of which in turn is indispensable for our brain development.[156] Severe iodine deficiency in the early years of life leads to cretinism, an extreme form of intellectual disability. At least 1.5 billion people worldwide live in iodine-deficient areas; at least 300 million do not develop their full intellectual potential.[157] Overall, this results in an average global IQ loss of 10 to 15 points. Thus, the undersupply of the world's population with iodine, along with that of aquatic omega-3 fatty acids, is one of the biggest single factors for a preventable mental developmental disorder.[158] Pregnant women, in particular, often have suboptimal concentrations of iodine, with tragic consequences for their offspring.[159] They should therefore always supplement with iodine if they want to enable their child to develop their full mental potential.[160] But even in the adult brain, hippocampal neurogenesis remains dependent throughout life on the effect of thyroid hormones and thus on an adequate supply of iodine, which is why a lack of iodine also leads to depression and a weakening of the frontal lobe battery.[161] The global undersupply is hard to comprehend—after all, the approximately 0.15 milligrams of iodine that an adult needs daily would be easy to cover with about five grams of iodized table salt in addition to a balanced diet.[162]

Lithium increases the release of hormonal growth factors that cause a capacity increase of the frontal lobe battery by means of hippocampal neurogenesis.[163] To achieve this effect, a small amount is sufficient. This is referred to as microdosing because it is about a thousand times less than that used in the treatment of bipolar disorder. To protect against depression and Alzheimer's and generally against chronic inflammatory processes, a daily intake of around 0.3 to 1 milligram is sufficient.[164] This corresponds roughly to the amount found in around 100 grams of maritime fish, which shows that lithium was also part of the brain-healthy diet of the fisherman and collector. But lithium doesn't just activate the hippocampal muscles' neurogenesis, it also blocks the vicious cycle of Alzheimer's toxin production set in motion by aggregated β-amyloid (see the box "Play wins, Alzheimer's loses" in Chapter 8) and activates cell-rejuvenating processes.[165] Due to this multipotency, it is the only monotherapeutic agent that can halt the Alzheimer's process when microdosed.[166] Due to this multipotency, it is the only monotherapeutic agent that can halt the Alzheimer's process when microdosed. Even in low concentrations, lithium extends our lifespan and is therefore considered essential.

The UGLY. Lithium, as we have seen, meets all the criteria of an essential trace element. However, it has not yet been officially recognized as such—at least in Germany—which prevents it from being sold as a dietary supplement or added to food in this country. It may be because lithium, even in very low concentrations, regulates various signaling molecules that are so important for our health[167] that the pharmaceutical industry wants to prevent people from being able to influence it themselves. Every year, large corporations invest billions of dollars in drug research just to develop products that influence these signaling pathways without side effects in the same way that microdosed lithium could—so far without success.[168]

Unfortunately, there are few natural sources of lithium other than marine fish. There is hardly any lithium to be found in local foods, and tap water rarely contains enough of it to meet requirements and achieve health effects. However, since even low concentrations offer a certain level of protection against depressive moods and even reduce the suicide rate, more and more scientists and doctors are calling for lithium to be added to our tap water.[169] But first the health authorities would have to recognize lithium as an essential trace element; then it would also be available inexpensively as a dietary supplement. This would make world health much better, but at the same time a huge future market would break away for our "health industry." The question that arises is obvious: What should take priority, our health today or corporate profits tomorrow?

Unless the government's response changes, I strongly recommend lithium-containing mineral or medicinal water to provide about one milligram of lithium per day.

Selenium helps detoxify heavy metals such as lead, cadmium, and mercury, which are known to cause brain damage.[170] Selenium is also an essential component of the selenoenzymes. These are endogenous proteins that require selenium to break down proinflammatory oxygen radicals, which are increasingly produced during high metabolic activity, for example, when neural stem cells proliferate. In selenium deficiency, the breakdown of these proinflammatory radicals is disrupted, and inflammatory reactions occur and thus immunological inhibition of the capacity expansion of the frontal lobe battery. Accordingly, a suboptimal blood concentration of selenium increases the risk of developing depression and Alzheimer's disease.[171]

Although a blood concentration of between 80 and 100 micrograms/liter is generally considered sufficient, an even lower cancer rate is only recorded at values of 120 to 150 micrograms/liter.[172] The daily requirement is around 60 to 70 micrograms, whereby one does not underdose or overdose.[173] Brazil nuts have the highest selenium concentration of up to 1900 micrograms/100 grams. There is a similar amount of selenium in sesame seeds, which is why one nut or three to four grams of sesame seeds a day is enough. Legumes, especially soy and lentils, as well as whole grains also have higher concentrations of selenium—but not if they were grown on the selenium-poor soils of central Europe.

Selenium and viral disease processes. Selenium is essential for defending against viral attacks.[174] Selenium deficiency, for example, increased mortality from HIV infections twentyfold.[175] Selenium deficiency, which is endemic in Europe, therefore also leads to an increased danger of viral infections. The selenium status in samples from surviving COVID-19 patients was significantly higher compared to those who died from it.[176] A sensible and above all necessary measure to reduce COVID mortality would therefore be to ensure that the population is adequately supplied with it.

Iron is found in both fish and plant foods, a much smaller amount than in meat, but sufficient. The ApoE4 gene helped our ancestors transport the valuable iron across the blood-brain barrier (which only selectively allows the iron needed in the brain to pass through). It was only much later that two additional ApoE variants emerged in Caucasian populations, ApoE2 and ApoE3, which have much lower iron transport capacity. Today, only about 15 percent of people are carriers of ApoE4, although there is good evidence that ApoE4 carriers also have an advantage in intelligence due to other functions of this original variant, but only under the living conditions of fishermen and collectors that were considered "species-appropriate" at the time.[177] In contrast to ApoE2 and ApoE3, modern but alien lifestyles make ApoE4 an accelerator of Alzheimer's dementia.[178] An explanation for this could be, from an evolutionary biology perspective, that thanks to ApoE4, enough of it arrived in the brain despite low iron intake. In individuals carrying the ApoE4 gene, a diet rich in meat and iron leads to increased levels of iron in neurons. However, an excess of iron causes cerebral overproduction of harmful oxygen radicals, which primarily impair hippocampal neurogenesis. For example, if there is too much iron, the frontal lobe battery capacity decreases, and the risk of Alzheimer's increases in the long term.[179] ApoE4 is therefore only an "Alzheimer's gene" if we eat a diet that is foreign to the species, i.e. meat-heavy. Conversely, ApoE4 carriers who are in the early stages of Alzheimer's respond particularly well to a change in their lifestyle: The disease process is not only stopped, but even reversed; a change even entails an enormous growth of the frontal lobe battery.[180]

As with all minerals, the law of the minimum also applies to iron. The maldistribution with too much dietary iron in the countries of the Global North is tragically "compensated" by too little in the countries of the Global South. Around two billion people suffer from iron deficiency, especially in Africa and Asia. Over 841,000 people die every year from anemia, the blood formation disorder caused by this.[181] A chronic deficiency of iron—like a chronic excess—is also harmful to brain development and causes long-term neurological impairments. In particular, the development of the frontal lobe battery is disrupted.[182] With a daily requirement of only 15 milligrams, a deficiency of iron would be easy to remedy, which would enable billions of people to live a better life.[183] Legumes and many other plant products would in this way provide sufficient iron, but are in short supply among the starving population. An excess of iron, from which billions of people worldwide also suffer, could just as easily be eliminated by giving up meat or at least significantly reducing meat consumption.

Essential amino acids are found in abundance in the protein of fish and seafood. However, they are also present in plant-based products in sufficient quantities to cover our daily needs. It is widely known that even such muscular creatures as elephants, cattle, and hippopotamuses eat a purely plant-based diet. Legumes such as soy, lentils, or peas have a comparable biological value in terms of their composition of essential protein building blocks such as fish, chicken eggs, or animal meat.

Based on its water and energy requirements, the microalga *Chlorella* supplies higher amounts of essential amino acids than any other plant species because it converts carbon dioxide and water into biomass very efficiently using solar energy. Well over half of its dry matter consists of valuable protein,[184] 20 percent each of which are complex carbohydrates and valuable fatty acids, such as the aquatic omega-3 fatty acid EPA. In addition, it contains high concentrations of chlorophyll, which also has an anti-inflammatory effect.[185] *Chlorella* also provides plenty of fiber, minerals, and many vitamins. I speculate that due to the limitation of agricultural land to meet the needs of an ever-growing global population, the future of food production will belong to microalgae such as *chlorella*. Another microalga called *Nannochloropsis*, or *Nanno* for short, could—produced under optimized culture conditions—even contain more than twice as much EPA as *chlorella*.[186] Their potential in protein production is just as considerable. For comparison: To produce one kilogram of essential amino acids from beef through factory farming, an average of about 150,000 liters of fresh water and 125 square meters of fertile land are required. In contrast, *Nanno* requires just 20 liters of water (a factor of 7,400 less) and only 1.5 square meters of land (a factor of 80 less)—and this does not even have to be fertile, since it is cultivated in incubators.[187] Like *Chlorella*, *Nanno* leaves no ecological footprint but in fact reduces the one created by our modern lifestyle.[188] This is because, in contrast to factory farming, which as the main protein source of the Western diet contributes to groundwater pollution with nitrate and agricultural overexploitation,[189] the waste products of microalgae can be used to improve soil quality naturally. In order for this sustainable type of protein production to be the future, a change in our lifestyles is necessary, which in turn requires a strong frontal lobe battery in the general population.

Vitamins are essential active ingredients and fulfill a wide variety of life functions. Many of these are abundant in fish and seafood, so it can be assumed that these served as the main source for our pescatarian ancestors, as with trace elements, essential amino acids, and aquatic omega-3 fatty acids. Exceptions are vitamins B_9 and K, which are also essential for our brain health, but are found almost exclusively in the plant-based part of the pescatarian diet.[190] In order to be supplied with all the vitamins that are contained in a "daily dose" of fish, you can also use plant or alternative sources.

Vitamin A (and other carotenes) can be found not only in fish, but also in kale, carrots, or spinach, as well as many other vegetables. Around 50 to 100 grams a day usually provides the required amount of around one milligram.[191] This is so little that it is difficult to understand why—as already discussed above—over a quarter of a billion children worldwide suffer from a serious vitamin A deficiency; some go blind and die early. But even a slight deficiency leads to disturbances in adult hippocampal neurogenesis and thus to a loss of frontal lobe battery capacity.[192]

For **Vitamin B_1 (thiamine)**, the daily requirement is also around one milligram. Legumes and whole-grain products are good sources.[193] However, if you only eat white flour products or refined rice, a thiamine deficiency can arise. Among other things, this leads to a capacity loss of the frontal lobe battery,[194] which applies to a deficiency of all B vitamins.

The daily requirement of around one milligram of **Vitamin B_2 (riboflavin)** can easily be met with whole-grain products and legumes.

The same applies to **Vitamin B_3 (niacin)**. Legumes are also ideal for the daily requirement of around fifteen milligrams. Peanuts in particular are rich in this vitamin.

We are able to produce **Vitamin B_4 (choline)** ourselves from the two essential amino acids lysine and methionine—therefore, strictly speaking, it is not a vitamin. Only essential substances that the body cannot produce itself but must absorb through food are referred to as such. However, to ensure that our organism has enough vitamin B_4, regardless of a possible weakening of its own synthesis, we should consume at least 0.3 grams of choline every day or lecithin (which is formed from choline) through food.[195] Just two to three chicken eggs (lecithin or Greek *lekithos* means egg yolk) would be enough to cover the daily requirement of this vitamin. Legumes are a vegan alternative, as are potatoes, many fruits and nuts, and almost all types of cabbage.[196]

We need around six milligrams of **Vitamin B$_5$ (pantothenic acid)** per day. This requirement can be met, for example, by one or two chicken eggs a day. Legumes and mushrooms are vegan options.[197]

The need for **Vitamin B$_6$ (pyridoxine)** is around one and a half milligrams daily. This amount can also be consumed in the form of legumes, whole-grain products, or nuts.[198]

We need around sixty micrograms of **Vitamin B$_7$ (biotin)** every day. A purely vegan diet with legumes, whole-grain products, and nuts would also be sufficient for this.[199]

Vitamin B$_{12}$ (cobalamin) is produced by soil bacteria. Herbivores consume vitamin B$_{12}$ through the roots of their food, which then reaches humans via the food chain as animal products such as eggs, fish, or meat. One or two chicken eggs cover the daily requirements if the chickens run freely and are allowed to eat grass with roots, which is very rarely the case. For a long time, products made from the microalgae *spirulina* were considered a vegan alternative. In fact, however, they only contain so-called pseudo vitamin B$_{12}$.[200] This is not only ineffective, but also competes with real vitamin B$_{12}$ when absorbed through the intestine, which can cause or worsen a vitamin B$_{12}$ deficiency.[201] Only *Chlorella* provides usable vitamins like B$_{12}$—around three micrograms, which is a sufficient amount—under certain culture conditions.[202] This amount is so small that our body actively absorbs this vitamin via a special transporter produced by the stomach lining. If the stomach lining is damaged (for example, by chronic inflammation), vitamin B$_{12}$ can only be absorbed passively—but then about a hundred times the amount must be supplied daily.[203] Since a B$_{12}$ deficiency is common and the resulting loss of frontal lobe battery capacity can be serious, I recommend regular medical checks of B$_{12}$ levels.[204]

Vitamin D$_3$ (cholecalciferol) is called the sunlight hormone because we can produce it through our skin when we are exposed to sufficient sunlight. Therefore, it is not a real vitamin, but the body's formation of it in the middle latitudes only succeeds in the summer months and in the far north not at all. Fish is also the main source here, because vitamin D$_3$, which is produced by plant plankton, accumulates through the marine food chain. For example, the Inuit cover their daily requirement of vitamin D$_3$ even without their own production. However—depending on the season and latitude—we are left with only one dietary supplement, which must be based on the blood values.

For optimal hippocampal neurogenesis or to prevent Alzheimer's, a blood concentration of around 100 nmol/liter (+/-30 percent) of 25-OH-D3 is necessary. This value of vitamin D circulating in the blood was determined by an international research team in a dementia study.[205] A moderate vitamin D deficiency of 25 to 50 nmol/liter increased the general dementia risk by 53 percent, and specifically for Alzheimer's by as much as 70 percent. The participants in the study with a value lower than 25 nmol/liter were twice as likely to develop Alzheimer's or vascular dementia.

The results of an extensive study with a quarter of a million Danes show a deficiency in the cardiovascular system with an increased risk of heart attack and stroke.[206] It not only confirmed the abovementioned necessary blood concentration of vitamin D, but also showed that too little is much more serious than too much. Only values well above 130 nmol/liter can be considered too high. So you should avoid overdosing, but also underdosing. A cancer prevention study also came to this conclusion, according to whose results much suffering could be prevented and many lives saved if sufficient vitamin D_3 were supplied.[207] According to the results of a meta-study by the DKFZ, sufficient supplementation with vitamin D_3 could prevent the death of 30,000 cancer patients over the age of fifty annually in Germany alone—which, however, is not happening so far.[208]

Low vitamin D levels thus increase the mortality rate from dementia, cardiovascular disease, cancer, and infectious diseases and reduce our life expectancy and, not least, the frontal lobe battery capacity. However, instead of the blood level of about 100 nmol/liter required for holistic protection and promotion of brain development, the annual average for German adults, for example, is only 45.6 nmol/liter, and in winter and spring it drops considerably lower.[209] For every nmol/liter deviation from the ideal value, at least 50 IU (international units) of vitamin D_3 are required daily, as a cancer prevention study found.[210] Therefore, in individual cases, dosages of several thousand IU daily may be necessary. However, to compensate for a deficiency, the German Nutrition Society (DGE) generally recommends a daily intake of only 800 IU of vitamin D_3 for adults. Although this is enough to prevent rickets (bone development disorder), it is far too little for everything else. How little interest there is in diagnosing a health-damaging vitamin D deficiency can also be seen from the fact that the examination is not a routine health insurance service. You must first show symptoms of osteoporosis in order for the costs of a 25OHD$_3$ measurement to be covered. But when you consider how much suffering could be eliminated with little effort, this practice is completely unacceptable.

The Gut-Brain Axis

Another part of our body plays an essential role in our brain health, which until a few years ago was not even considered as such. But for several years now, we have known that we need the microbial diversity of our large intestine, the microbiome, as a gigantic bioreactor for life.[211] The microbiome belongs to our digestive system and produces some B vitamins and vitamin K. But it is also hormonally active and crucial for the development and function of our entire immune systems—as well as for our brain.[212] If the composition of the well over a thousand different bacterial populations and viral strains of the microbiome is changed in animal experiments, some brain functions and even behaviors also change. The genetically diverse microbiome is essential for adult hippocampal neurogenesis and thus for learning ability as well as social behavior.[213] And since the microbiome already exerts such an influence on brain function in animals, we can assume that a similarly close symbiosis must have developed between our brain and the microbiome over a very long period of evolutionary history.

The optimal composition of the human microbiome is therefore very likely based on the diet of prehistoric fishermen and gatherers. Indeed, the Yanomami people, who have lived a nearly untouched Paleolithic life in the Venezuelan Amazon for about eleven thousand years, show the greatest microbial diversity discovered so far.[214] Due to the Western diet, which now dominates in almost all countries of the world, there is, however, a serious loss of microbial diversity in the microbiome.[215] Fast food or heavily processed foods with too many simple sugars, unhealthy fats, and especially too little fiber are problematic for our microbiome. For example, bacterial strains that help digest complex sugars have now been completely lost.[216] Similarly, the often overly strict hygiene rules, generous use of antibiotics, and use of medications such as antacids (for heartburn) or metformin (for type 2 diabetes) reduce the genetic diversity of the microbiome in people in industrialized countries.[217] Thus, a veritable extinction of species is taking place within us, which threatens not only our microbiome but also our health—with fatal consequences: According to the results of a global study, around 11 million people died in 2017 alone due to poor nutrition, most of them due to a lack of fiber-rich foods such as whole grains or fruits.[218] Leading microbiome researchers write accordingly in the journal *Science*: "We owe future generations the microbes that have colonized our ancestors for at least 200,000 years of human evolution."[219]

Autocannibalism. Fiber is the food basis for intestinal bacteria and thus at the same time the basis for the complexity of the microbiome. If we eat a low-fiber diet, our intestinal inhabitants look for other food and instead eat into the protective mucous layer of our digestive tract. This has serious consequences because this barrier protects us from invasive and therefore highly dangerous microbes.[220] A healthy, fiber-rich diet is therefore important to keep our intestinal bacteria from eating us.

If we eat a diet rich in complex sugars and health-promoting fats and high in fiber—whole-grain products, nuts, and legumes, as well as lots of fruit and vegetables—our microbiome, digestion, and supply of vital nutrients improve. You should increase your fiber intake slowly, over a few weeks, so that the microbiome can adapt. Then it promotes the function of our immune system and, last but not least, adult hippocampal neurogenesis and thus our mental fitness.

In fact, you eat smarter with grains: Scientists fed wild-caught great tits either insects (low in fiber) or grains (high in fiber).[221] After just a few weeks, the vegan, high-fiber diet not only ensured a greater diversity of intestinal bacteria, but also made the birds smarter. So they were more creative in solving problems that researchers made them face while foraging. The scientists therefore speculate that a more complex diet and a correspondingly more complex microbiome are responsible for the diversity of behavior observed in wild populations.

Our modern, species-foreign diet not only reduces our vital biodiversity in the microbiome, it also ensures that our organism is flooded from the gut with both non-specific activators of the immune system, which trigger inflammatory reactions in the brain among other things, and with protein complexes that are structurally similar to the Alzheimer's toxin and also damage the brain.[222] Both mechanisms inhibit adult hippocampal neurogenesis and at the same time promote neurodegenerative processes, which increases the risk of developing Alzheimer's in the long term.[223] But even in the short term, existing anxiety symptoms intensify and the depression rate increases.[224] There are indications that even ADHD could be co-caused by a disturbed microbiome.[225]

You Are What You Eat

The more complex the food, the more complex the microbiome and the more complex the behavioral repertoire. Thus, people who eat a modern species-appropriate diet, a quantitatively and qualitatively healthy diet, have a frontal lobe battery with a larger capacity. If, on the other hand, we eat a simplistic diet, we are more likely to be so in our thoughts and actions. Deficiencies in our food are ultimately the cause of deficiencies in our thinking skills. This shows the full extent of the problem that our modern diet is causing for our individual behavioral repertoire, but also that of the global community as a whole.

CHAPTER 12

HUMAN—THE ANIMAL OF MOTION/ANIMAL OF MOVEMENT

Life is like riding a bike. To keep your balance you must keep moving.
—Albert Einstein (1879–1955)

Agile on Foot—Agile in the Head

My schoolmates and I are once again sitting in the classroom sweating. I have to dry my face and smile into the tissue. It was a fair fight and everyone won: There are no losers in the game because everyone has fun. Only the crumpled and bruised "Schoki" Tetra Pak [chocolate milk box] doesn't make a good impression and is hardly recognizable. As at every break, the container of a quarter liter of milk with heavily sweetened cocoa served as a "football" and four more cartons as "goal posts." We boys come from five different villages, and almost all of us play in our respective local football clubs. For us, the break bell is not just an opportunity "chocolate ball game," but as if it was the kickoff to the next local derby. Today, almost half a century later, when I think back to this wonderful time, I ask myself how I manage to spend hours sitting and writing in front of a computer screen, without any interruption from exercise, without becoming permanently ill. The answer: I can't do it. I know it, and I urgently need to change something.

One might assume that the intense physical activity during the school break may have tired us boys and therefore made us less able to pay attention to the lessons that followed. But I never had that impression, and as we now know, the exact opposite was most likely the case: Even short-term physical activities have a particularly positive

167

effect on those executive functions of the brain that are closely related to learning performance. Studies by the Max Planck Institute for Human Development have shown
that children—and adults as well—solve learning tasks significantly better when they
move on a treadmill than when they sit on a chair.[1] The positive effect of physical
activity on the mind is all the greater the more complicated the mental demands are.
This improvement in cognitive performance only occurred when the participants were
allowed to set the treadmill's pace themselves and were not forced to a specific pace,
which caused stress and blocked thinking.

Interestingly, material can be learned more efficiently not only during self-determined movement, but also for some time afterward, as studies show.[2] Even a thirty-
minute athletic exertion promoted the performance of working memory and improved
both cognitive flexibility and attention in thirteen to fourteen-year-old students in
seventh grade in Germany.[3] If one compares the cognitive test results of children who
were previously physically active for thirty minutes with those of children who only
watched television while sitting during the same time, the former far surpass the latter.[4]
According to further studies, they also react faster and more accurately to a variety of
cognitive tasks, even if they were only physically active for a short time.[5] The evidence
is therefore clear: Physical activity and fitness improve academic performance, with not
only physical education lessons but also break times during the school day being of
great importance. At the center of these effects is the frontal lobe battery. Sports scientist Christian Buchmann, who led a similar study, noted accordingly: "Exercise doesn't
make children smarter, but it does make them more absorbent and more capable of
concentrating in the sense of successful learning processes."[6]

When we are physically active, a few things happen in our brain. First, exercise
causes an immediate increase in blood flow to the brain. Even with a low power of just
25 watts, such as when going for a walk, there is a more than 13 percent increase in
the oxygen and energy supply to the brain.[7] At around 100 watts, as achieved during
casual jogging, cerebral blood flow even increases by around 25 percent above the value
when sitting. So you don't have to achieve top Olympic performances to make your
brain more efficient.

But our brain is also more capable of remembering in the long term if we "move"
it from A to B more often on our own. Here too, the explanation can be found in our
prehistoric past. Early in the evolutionary history of humans, at the level of our animal
ancestors, a well-functioning memory of experience was vital, for example, in order
to remember where, when, and what was experienced and whether it felt either good
(new food source discovered) or bad (traces of a dangerous predator discovered). The
answers to these where, when, what and how questions are exactly those that the frontal lobe battery had to remember (as part of the archicortex, even birds already have a
corresponding equivalent). It thus functioned simultaneously as a memory of time and
place, in order to return home safely and on time. This was essential because anyone

who couldn't make it back to the cave or nesting site even once was endangering the successful transmission of their genetic program. It therefore seems logical that, in the course of evolution, genetic mechanisms developed in more complex brains, and ultimately also in that of humans, that improve hippocampal memory capacity in the long term through physical exercise.[8] Countless studies have now identified over a dozen hormonal signaling substances that are released directly or indirectly through muscle work, which then not only strengthen the musculoskeletal and cardiovascular systems but also improve mental abilities in the long term.

The office worker's dilemma. A major genetically determined difficulty that modern humans find themselves in is that throughout the formative part of their developmental history, they only identified as "The Animal of Movement" and at the same time were programmed to save energy, because food was not always abundant in its evolutionary history. So they instinctively moved as much as necessary and as little as possible. Until not so long ago, food security was only guaranteed through physical activity. But now most people earn their daily bread sitting down. Even farmers usually only move the levers in the driver's cab of their huge tractors and harvesters; the rest is usually done—at least here in Germany— by harvest helpers from poorer countries. And last but not least, our children are raised to pursue careers with little movement. Although this is economically compliant, it is by no means appropriate to the species.

For example, if we take the stairs instead of the elevator to get to a higher floor, our muscles use oxygen to "burn" carbohydrates to provide the extra energy needed to do so. However, since breathing and the cardiovascular system reactions are always somewhat delayed, the oxygen content in the muscles initially drops a little. This delay creates an oxygen debt and is the reason why our heart still beats a little faster and we have to breathe a little harder for a while after we have already reached the top—the oxygen debt is then compensated for. However, our organism uses the initial drop in oxygen concentration as information to improve its future performance. Sensors in the inner lining (endothelium) of the blood vessels register the local lack of oxygen and produce vascular (concerning the blood vessels) endothelial growth factors (VEGFs). VEGFs cause new blood vessels to sprout in the affected muscles so that they receive better blood flow and oxygen next time. Physical fitness increases. However, VEGFs also signal the frontal lobe battery to form more nerve cells.[9] Even with light physical activity, there is also an increased formation of new blood vessels in the brain, which then improves cerebral blood flow even at rest.[10] Mental fitness increases.

Exercise also strengthens our immune system, with some of the immunological messengers (cytokines) released during exercise also contributing to the growth of the frontal lobe battery.[11] Even our muscles produce hormones—as we have seen—as soon as we use them. For example, they then release irisin, which regulates fat metabolism and at the same time increases growth of the hippocampus. Physical activity also increases the production of growth hormone (GH), which is released with a delay at the beginning of deep sleep at night and promotes regeneration and muscle growth. In addition, GH is a very potent activator of adult hippocampal neurogenesis.[12] The larger the muscle groups you allow to work and the longer the activity, the more GH is released and the better regeneration and neurogenesis. These effects are the reason why exercise is good for the whole body. Adiponectin, which we already learned about in the previous chapter as a fat cell hormone that is beneficial for neurogenesis, is released more during physical exertion, has an anti-inflammatory effect, and protects against, among other things, arteriosclerosis and type 2 diabetes.[13]

If a large part of our muscles works a little longer and more intensively, the oxygen demand increases not only at the place where the activity occurs, but throughout the body, which is then also registered by the kidneys. These also produce erythropoietin (EPO for short), a hormone that has many functions. Among other things, it ensures that more red blood cells are produced in the bone marrow. As a result, the blood can carry more oxygen next time, increasing physical and mental performance. But EPO, like VEGFs, also provides a strong growth signal for the hippocampus.[14] This is one reason why the pharmaceutical industry wants to develop EPO as a therapeutic agent against Alzheimer's, but will probably fail in doing so, because a natural combination of many messenger substances and hormones is necessary to protect us from mental decline (and many other minima would also have to be corrected, as you now know). Only when we are physically active does the hippocampus benefit from not just one, but a whole armada of growth factors. A big advantage of adult hippocampal neurogenesis activated by physical exercise is that it is free of undesirable side effects. Not a single hormone is overdosed in this natural way. However, if you try to achieve the same goal of improved physical and mental fitness without exercise with a high-dose monotherapeutic agent, the opposite happens. Artificially increasing the concentration of such hormones, as we already know from the doping scene, is extremely unhealthy.[15]

Runner's high—or how nature encourages us to move. Those who exercise are rewarded with the release of endorphins, the body's own "opiates." The endorphins are small proteins and the reason why it is usually fun to be physically active and why you like to explore your limits. They are responsible for the so-called "flow." The release of the feel-good messenger substances serotonin and dopamine, which also stimulate hippocampal neurogenesis, also contributes to this.[16]

This is experienced as a feeling of happiness when you are completely immersed in an activity and often occurs in endurance sports. However, all of these messenger substances are also necessary for adult hippocampal neurogenesis increased by physical activity to take place at all, which means that artificial administration of EPO or GH is not sufficient to simulate the mental effects of exercise.[17]

The multiplication of blood vessels (VEGF) and oxygen transporters (EPO) while simultaneously strengthening the muscles (GH) make climbing stairs (or any other physical work undertaken) a little easier the next time, and this facilitation is called the training effect. However, in order to get your brain on a growth path through physical activity, you have to not exercise intensively. For this purpose, exercise in the aerobic range is sufficient if the intensity of physical activity and breathing are in harmony. Sufficient oxygen is supplied to efficiently convert carbohydrates and fatty acids into kinetic energy. From an evolutionary biology perspective, it makes sense that aerobic activities are sufficient to stimulate brain growth: Our ancestors also most likely traveled at a leisurely pace, as neither fishing nor gathering works particularly well in a sprint. If you want to determine whether you are still in the aerobic range during your sporting activity, you do not need any electronic aids. As long as you can get just enough air through your nose during an activity with your mouth closed, you are at around 70 percent of your maximum heart rate and therefore in the optimal aerobic training range. Accordingly, it has been shown in men that just forty minutes of physical activity of this intensity causes an increase in the already known brain-derived neurotrophic factor (BDNF) of around 32 percent.[18] Although BDNF is produced in the brain, it also passes from there into the blood, so that the concentration in the brain is closely related to that in the blood, where it can be easily determined. In contrast, the BDNF concentration in the blood and thus also in the brain of the sitting control subjects decreased by 13 percent within the same time. Longer training sessions had a greater positive effect; longer sitting times had a negative effect. The latter is highly problematic because a chronic BDNF deficiency accelerates the aging process. Thus, it could be shown that low BDNF levels are causally linked to the development of Alzheimer's disease, since the release of this neuronal growth factor is indispensable for the preservation of our brain mass.[19]

Exercise beats Alzheimer's gene. In an eighteen-month study, older people of average fitness who carried the ApoE4 gene variant in their genome were observed for their exercise behavior. ApoE4 is commonly viewed as an Alzheimer's gene, although, as discussed in the previous chapter, it supports brain health with a species-appropriate lifestyle that includes exercise. Those seniors who were not very physically active lost around 3 percent of their hippocampus volume during the observation period, which brought them a lot closer to Alzheimer's disease.[20] However, if they were physically active, the researchers were unable to detect any loss of volume. In another study it was found—and this is not surprising—that walking just over 3 kilometers a day, compared to just 400 meters, which corresponds to our alien normality, is enough to reduce the risk of Alzheimer's disease by half.[21]

Creativity Is Thinking in Motion

"The hippocampus has only recently been linked to creativity and imagination," explains Dr. Wendy Suzuki, professor of neuroscience and psychology at New York University's Center for Neural Science, "because imagination is about taking the things you have in your memory and putting them together in a new way."[22] Creative intelligence is thus the ability to recognize new connections and retain and use them in a meaningful context over the long term. The hippocampus or its frontal lobe battery is predestined for this. It enables us to quickly learn complex, emotionally significant connections. This is also referred to as associative learning.

Experiences only have meaning in relation to other experiences, and our perception is also significantly influenced by previous experiences. The American writer and social and cultural anthropologist Carlos Castaneda (1925–1998) wrote: "We perceive. This is an undisputed fact. However, what we perceive is not an equal fact, because we learn what we have to perceive."

If we train our children to sit still from an early age so that they supposedly learn better, the resulting lack of physical activity overrides the very genetic mechanisms that would make associative learning easier for them. Nine- to ten-year-old study participants who exercised a lot and were physically fitter were also better at associative remembering: They grasped connections more quickly and retained them more easily.[23] The physically less fit children were able to remember facts just as well (this is possible without the hippocampus), but their ability to remember experiences was limited and in proportion to their lack of movement or the relative reduction in size of their hippocampi. The researchers discovered a significant three-way relationship: Physical performance, mental fitness, and the volume or frontal lobe battery capacity are positively related to each other. The conclusion of the scientists involved in this study: "These results are the first to indicate that aerobic fitness influences the structure and function of the infant human brain."

It's Never Too Late for an Eventful Childhood

BDNF blood levels of expectant mothers, even *before* birth, significantly influence the brain development of their children.[24] Animal experiments also found that in pregnant rats, the memory capacity of their later offspring is related to their amount of exercise, with positive consequences for their memory.[25] Therefore, pregnant women should remain physically active and also take other measures that activate BDNF for them and their growing fetus. All elements of the "formula for a high-capacity frontal lobe battery" are capable of this and necessary: In addition to social engagement, this includes nightly fasting, avoiding refined sugar, and an adequate intake of aquatic omega-3 fatty acids. In addition, exposure to sunlight, as in outdoor exercise, can further increase BDNF release.[26] The path to greater frontal lobe strength thus begins before birth and continues throughout life.

> **Dynamic genetic material.** Researchers have found that exercise has a positive (epigenetic) effect on our genetic program. In athletic people, several hundred genes that optimize energy metabolism, promote muscle building, and protect against oxidative stress are preferentially read.[27] As a result of this exercise-induced gene programming, muscle cells can, for example, absorb sugar from the blood more efficiently, which protects against type 2 diabetes and should be used therapeutically for all forms of diabetes.[28] Moreover, there is even an epigenetic inhibition of the production and release of proinflammatory regulators, which protects against chronic inflammation.[29] Since we know that not only genetic but also epigenetic changes (beneficial or harmful to well-being) in our genome are inherited, the question for future parents, a purely personal one, "Sport or no sport?" is of cross-generational importance.

Physically active children are less likely to become depressed, as was discovered in the Trondheim Early Secure Study.[30] Conversely, depressive symptoms at the age of eight to ten are related to lower physical activity, whereby, according to the study results, the lack of exercise is the cause of the development of depression and not just its consequence. The study also showed that children who were particularly physically active at the age of six maintained their urge to exercise throughout the entire four years of observation. However, even the more lethargic children remained true to their behavior. Apparently, not only adults but also children find it difficult to change their movement behavior once it has become established. This could certainly be explained by a correspondingly adjusted strength (a lot of exercise) or weakness (little exercise) of the frontal lobe battery capacity. In fact, another study showed that children aged nine to eleven years had a larger frontal lobe battery if they spent their free time playing sports rather than engaging in art or music.[31] As expected, the risk of developing depression also decreased in this study as the hippocampus grew. This is not to say that you should neglect art or music—on the contrary. But you shouldn't neglect physical activity in addition to a predominantly sedentary job. As with everything in life, it's all about balance.

Since a growing hippocampus keeps us young, at least mentally, it is never too late for an "active childhood." In my opinion, the best indication that an increase in frontal lobe battery capacity is actually possible even at an older age was published in 2011 by the US Academy of Sciences.[32] For their study, the researchers selected 120 seniors with an average age of about seventy who had age-appropriate mental health, i.e., were "normal," and who had not walked more than thirty minutes per week in the last six months before the start of the study, which is unfortunately also "normal" for most elderly residents of social housing. Then the participants were randomly divided into two groups of equal size. One group was asked to go for a walk for about forty minutes every day. The other group only did gymnastic stretching exercises while sitting. In total, the participants either hiked or stretched for a year. All participants' brains were volumetrically measured using imaging techniques before, in the middle, and at the end of the study. The results were clear: In the walkers, the hippocampus in particular grew by an average of around 2 percent over the course of one year, with the increase in volume increasing the fitter you became. The same applied to the performance of episodic memory and psychological resilience. In the gymnastics group, however, the hippocampus shrank by around 1.4 percent. This is slightly more loss of frontal lobe capacity than the population average of about 0.8 to 1.0 percent. But with less than thirty active minutes per week, or just over four minutes on their feet per day, the gymnastics group was also slightly below the exercise average for the general population—but only slightly.

Based on the clear results of a meta-study, US scientists from the department of psychology at the University of Pittsburgh concluded that "higher cardiorespiratory fitness levels are routinely associated with greater gray matter volume in the prefrontal cortex and hippocampus are associated, and less consistently in other regions."[33] This applies in old age, but also in youth, and explains the results of an international study, according to which "higher cardiorespiratory fitness is important for the health of children and adolescents and also to improve school performance."[34]

The many functions of the numerous messenger substances that are released during physical activity are the reason why exercise improves cognitive processes and memory, but also strengthens the immune system and physical and psychological resilience, has analgesic (pain-relieving) and antidepressant effects, and last but not least creates a feeling of well-being. It is therefore difficult to understand why exercise is no longer encouraged and why even children are allowed to move less and less.[35]

Sitting Is the New Smoking

Lack of exercise is one of the most important preventable causes of death worldwide. According to the WHO, adults are getting enough exercise if they spend more than 150 minutes on their feet per week, which simply means not sitting. Or if you do more than 75 minutes of moderate-intensity aerobic exercise per week. That would be about 22 minutes of non-sedentary activity or just under 11 minutes of physical activity per day. One can only speculate why the WHO, which has committed itself to nothing less than improving world health, specifies such low target values that are too low to prevent a chronic loss of frontal lobe battery capacity, let alone significantly stimulate its growth. But even these WHO standards are far too low; not even 42 percent of German adults meet them. These frightening figures come from 2016. Two years earlier it was still 54 percent—it can be assumed that the downward trend has continued.[36]

Exercise is better than antidepressants. Regular physical activity not only strengthens mental resilience and thus reduces the likelihood of slipping into depression or suffering burnout, it also reduces mortality in severe depression.[37] Nevertheless, it is still not used sufficiently in clinical practice, possibly due to some widespread misperceptions or outdated teachings. It is still not clear to most doctors and therapists that adult hippocampal neurogenesis must be the focus of causal prevention and treatment of depression and that exercise promotes this in many ways. Only then would a lack of exercise be recognized as a cause of depression and thus exercise as a causal prevention and therapy—which is just as effective acutely as any antidepressant and even works better in the long term or reduces the likelihood of a relapse more.[38] In addition, it is often wrongly assumed that that many patients would not follow an exercise program even if doctors prescribed one. This can be achieved in a large number of cases if the doctor has time for a motivating conversation and actually prescribes the exercise.[39] The assumption that only strenuous exercise is effective is—as you now know—incorrect.

The fact that twenty-two minutes of light or eleven minutes of moderate physical exertion per day is not sufficient to reduce the risk of cancer, for example, is shown by a long-term study of over eight thousand seniors, on average about seventy years of age.[40] The scientists found that the risk of dying from cancer decreased by 8 percent for every half hour of light activity. The same amount of more intense activity even reduced it by 31 percent compared to a purely sedentary job. With these results, the authors show "that the total volume of sedentary behavior is a likely risk factor for cancer mortality" and conclude "that adults should sit less and move more to promote longevity." Exercise lowers not only the probability of developing certain types of cancer, but also of diabetes, heart attack, or stroke, as a meta-study found.[41] While the WHO calls for an additional expenditure of only about 600 calories per week through exercise (which corresponds to about seventy-five minutes of aerobic activity), according to this study, a daily (and not just weekly) additional calorie expenditure of this magnitude was necessary for a convincing reduction in the occurrence of all these life-threatening diseases.

The UGLY: The trend toward more and more physical inactivity continues unabated despite all the knowledge about the long-term health-endangering consequences. The *ÄrzteZeitung* reported: "Rich countries like Germany are among the drivers of the negative development. In these countries, the proportion of physically inactive people in the population has increased from an average of 31 percent (2001) to 37 percent (2016)."[42] The industry thrives on simplifying our lives. Everything has to be motorized. The Federal Ministry of Transport, under Andreas Scheuer, has also promoted environmentally harmful electric scooters from the beginning, under the motto of the "last mile" from the subway and S-Bahn stations or from bus stops to home or work.[43] The main thing is that nobody has to move more, which means the frontal lobe battery shrinks so that the economy grows. Humans are animals of movement; sitting shrinks the brain, makes you sick, and kills prematurely. If you want to be mentally fit and live long, physical activity is the best medicine.

Outrunning Age

If we live in a species-appropriate way, the frontal lobe battery does not lose any capacity or function. Since the ability of our entire brain to regenerate remains intact well into old age, the exciting question arises as to what actually limits our lifespan. After all, you are only clinically considered dead when your brain gives up "its ghost."

Most people die from failure of their cardiovascular system, which then indirectly leads to brain failure. As a rule, our lives are limited not by our mental performance, but by our physical performance, and this can be measured. To do this, scientists determined the volume of oxygen (VO_2 max) that we can absorb within one minute during maximum physical exertion. The VO_2 max is limited by the function of the lungs (maximum oxygen uptake capacity), the blood (maximum oxygen transport capacity), the cardiovascular system (maximum oxygen distribution through the maximum pumping capacity of the heart), and the muscles (maximum oxygen processing). If just one of these systems loses performance, the VO_2 max and thus overall physical performance decrease—a chain is only as strong as its weakest link. VO_2 max is therefore a simple and reliable measure of a person's physical fitness. However, instead of the absolute VO_2 max (measured in milliliters per minute, ml/min), the relative VO_2 max is calculated and used in studies. By dividing the VO_2 max value by body weight (ml/min/kg), the values become comparable between individuals, but also within the individual life course. With the help of the relative VO_2 max determination, the vital question can now be answered as to how a person's performance changes over the course of life and which factors influence it. A first (but fortunately not entirely correct) answer was provided by cross-sectional studies in which the VO_2 max of many test subjects of different ages were examined at a single point in time.[44] In this simple and quick way, it was found that twenty-five-year-olds with a relative VO_2 max of an average of 47.7 ml/min/kg had the highest value. This was almost halved in seventy-five-year-olds, with a value of 25.5. Over a period of five decades, this resulted in an average loss in physical fitness of around 1 percent per year. This degradation rate also corresponds to that of the frontal lobe battery in the normal population; body and mind break down in lockstep, so to speak. But as we now know, the loss of frontal lobe battery capacity is completely unnatural—not only can it be stopped by exercise and the other factors of the formula for a high-capacity frontal lobe battery, but an increase is even possible.

Does our body take a drastic special path in this case? Initial indications that physical performance is also maintained for much longer with a species-appropriate lifestyle than these results of the cross-sectional studies suggested were provided by a series of longitudinal studies.[45] In these, the VO_2 max of test subjects was examined repeatedly over many decades, whereby their individual performance could be directly correlated to their respective lifestyle, and a relationship between the two could be set. These very time-consuming studies found that people who exercised regularly throughout their lives only lost about half a percent of physical performance each year from age twenty-five onward, whereas people who lived a life completely free of sports lost almost 2 percent per year. Without any exercise, people aged about twice as fast as the average population (at least a small part of which regularly exercises) and about four times faster than the athletic subjects in the study. According to the results of another study, athletes who did not only a little sport, but even did competitive sport, lost an average of only 0.17 percent of their VO_2 max annually between the fifth and sixth decades of life, over ten times less than sedentary people.[46]

From a purely physical perspective, based on the results of these longitudinal studies, a lifespan of 120 years is possible.

If you have spent most of your life sitting, then I can reassure you that it is never too late to start exercising. Even fifty-year-olds can regain their lost physical fitness after thirty years of athletic abstinence.[47] Thus, it is clear that people who have been physically inactive for a long period of their lives have a huge untapped life potential that can be released by switching to more exercise. We just can't stop being physically active. And since body and mind are closely linked, such an almost biblical age can even be achieved with full mental capacity. Our inner life clock also points to this connection; it consists of telomeres.

The telomere is the repetitive part at the respective ends of our genome. Telomeres protect that large information or gene-bearing intermediate piece. This protection is necessary because in every cell division, when the genetic material is duplicated, a small piece is lost at the ends due to the reproductive mechanism. If the telomeres become too short, further division of the cell is no longer possible. Already in the middle of the last century, Leonard Hayflick and Paul Moorhead discovered that a human fetal connective tissue cell in culture can only divide about fifty times and described the progressive loss of ability to divide as "cellular senescence" (from the Latin *senescere*, "to grow old.") This "Hayflick limit" could explain why our lifespan is limited.[48] The increasing shortening of telomeres in the course of life is also closely related to the probability of developing cancer and suffering a heart attack or stroke.[49]

Omnipotent stem cells, from which our germ cells also emerge, can activate telomerase. This is an enzyme that ensures that the genetic material of egg and sperm cells, which ultimately have to become an entire organism (which means many cell divisions), always have the maximum telomere length. Evolution over millions of generations is only possible thanks to telomerase. Recent findings now prove that even ordinary body cells are capable of this—provided a species-appropriate lifestyle.[50] Thus, it was first discovered in observational studies (in which only a connection is observed), and later also in controlled intervention studies (where the connection is tested for its cause) that, among other things, physical activity leads to longer telomeres, i.e., activates telomerase.[51] It is enough to interrupt sitting every now and then to lengthen the telomeres somewhat or to rewind the life clock a bit.

This all sounds very simple, but it is not. The executive center of our brain decides whether we move ourselves or whether we prefer to remain seated or whether we move primarily with motorized assistance. A self-propelled brain is more mentally agile and therefore more capable of making the right life decisions. Moving yourself is a vital decision.

CHAPTER 13

OUR DAILY POISON

Excessiveness is poison for any kind of self-knowledge.
—Ernst Ferstl, Austrian poet

No Right to Clean Air

I was a chain smoker. For the first fifteen years of my life, I inhaled the smoke of several packs of cigarettes every day. Strictly speaking, the vice began when I was born; my father and grandfather made sure that the air in our home was always saturated with cigarette smoke. I remember how watching TV together was only possible with watery eyes and a scratchy throat. You constantly had the acrid smell of tar in your nose and mouth. Often, the cigarettes would be burning simultaneously in different ashtrays because my father, in his work-related stress, would forget that he had already lit a cigarette in another room. Added to this were the unfiltered cigarettes of my grandfather, who visited us daily to fulfill his original career aspiration as a pensioner, namely to be a teacher, and now tried to teach me what he himself was good at: English, chess, and piano. I also remember that my mother had to take down the curtains weekly to wash out the ugly yellow. Unfortunately, it didn't occur to anyone that our lungs were also yellowing. Nor does it say that the "exhaust fumes" are carcinogenic or that they can damage a child's brain. Because I knew nothing else, the bad air was a normality that I only began to question in my puberty.

If I were to get lung cancer one day, I know who I would blame. Not my father or my grandfather. That would be obvious, but it would be wrong. After all, they too were just victims. When my grandfather started smoking in his youth, tobacco smoke was considered to be good for health: "Smoking makes you slim, healthy and happy!" was the tenor of the advertising slogans at the time.[1] But even in the 1950s, when my father started smoking, it was still considered "cool" to hold a smoking cigarette in your hand.

It was even believed that the smoke killed pathogens, which was why smoking was recommended if pneumonia was suspected. Perhaps my desire to constantly question medical doctrine stems from my personal experiences regarding these circumstances. Even when they are outdated, they remain persistent in people's minds for a long time, especially if they are good for business. See my books *The Alzheimer's Lie* (Conventional wisdom: Alzheimer's is inevitable, you just have to get old enough) or algae oil (doctrine: linseed oil is enough to provide yourself with sufficient essential omega-3 fatty acids).

However, rumors were already circulating at that time that the burning sticks could cause cancer and heart attacks and were therefore extremely unhealthy. Nevertheless, for many years to come, the Marlboro Man on his mustang was the epitome of "male" freedom, which more and more women wanted to "enjoy." The first Marlboro Man and at least three others are rumored to have died as a result of smoking.[2] Nevertheless, the findings regarding the harmfulness of smoking were dismissed as talk for a long time and ignored. The tobacco industry and its lobbyists cleverly obscured the clear view of the facts. They not only denied their knowledge of the health-endangering effects of their products, they even claimed the opposite in public hearings.[3] According to the WHO, the tobacco pandemic is still "one of the greatest threats to public health the world has ever faced. More than 8 million people around the world die from it every year."[4] That is twice as many as from and with COVID in 2020. According to studies by the WHO, Germany is at the bottom of the list when it comes to smoking prevention policy.[5] Children in particular are still not protected by appropriate laws. For example, smoking in vehicles is allowed when children are present, even though pediatricians have long called for a ban.[6] However, such a ban does not seem to be politically desired, because here too the comparison to COVID shows that if you really want to pass laws, you have to do it at record speed.

Tobacco smoke, like a severe COVID infection, leads to the destruction of the lungs—more slowly, but with a much higher probability. According to the WHO, about 50 percent die as a result of smoking.[7] For comparison, only around 0.2 to 0.3 percent of those infected die from COVID.[8] One argument from smokers against the dangers of the toxins contained in tobacco smoke and against the fact that they can cause lung cancer is the often-heard statement that nonsmokers also get lung cancer. The cynical thing about this statement is that passive smoking significantly increases the risk of lung cancer.[9] According to the WHO, more than 1.2 million people worldwide die each year from tobacco without consuming it themselves—that's more than 15 percent of all deaths from tobacco smoke.[10] About 165,000 of them are children, and around 1.8 billion nonsmokers are also at risk, which was confirmed by a study published in *The Lancet* by the Institute for Environmental Medicine at the Swedish Karolinska University in Stockholm in 2011.[11]

Even passive smoking, a lower concentration of cigarette smoke, inhibits the intellectual development of our children and impairs their ability to learn.

The children of mothers who smoke a pack of cigarettes per day during pregnancy, which roughly corresponds to average consumption,[12] have an IQ that is on average 2.87 points lower than children born to nonsmoking mothers.[13]

- The reason for this could be that smoking (passive or active) inhibits the development of the frontal lobe and its battery.[14] According to Edythe D. London, professor at the Brain Research Institute at the University of California, Los Angeles, such an effect can "influence young people's ability to make sound decisions regarding their well-being, and that includes the decision to quit smoking."[15]
- It is estimated that over 21.9 million children in the United States alone have reading difficulties directly attributable to secondhand smoke. Higher exposure to passive smoking also leads to correspondingly greater deficits, including in mathematics and spatial imagination.[16]
- Children develop a nicotine addiction from passive smoking. Children of smokers who were regularly exposed to tobacco smoke at home or in the car show symptoms of nicotine addiction.[17] According to Dr. Jennifer O'Loughlin, lead author of the study and professor in the department of social and preventive medicine at the Canadian Université de Montréal, these symptoms include "depressed moods, sleep disorders, irritability, anxiety, restlessness, difficulty concentrating and increased appetite."[18]
- Nicotine addiction through passive smoking could be one reason why children growing up in smoking households are more likely to become smokers themselves than children in households where there is no smoking. In addition, in a smoker's household, existing cigarettes make it easier to try them out. In addition, parents are often a role model for children, even in a negative sense.
- Adolescent smokers have an underactive frontal lobe, as another study found: "The greater the nicotine addiction of a teenager was, the less active the prefrontal cortex [or frontal lobe]."[19]

Cigarette smoking leads to long-term thinning of the cerebral cortex, especially in the area of the frontal lobe, but also in the regions that are primarily affected by Alzheimer's.[20]

Nicotine itself directly inhibits hippocampal neurogenesis and is also neurotoxic.[21] This accelerates the already unnatural capacity loss of the frontal lobe battery in our society,[22] and the ability for System II thinking is restricted, although this is precisely what would be necessary to change unhealthy behaviors—unfortunately, also to quit smoking.

Of course, if you grew up as a child in a household where there was constant smoking, you are not mentally impaired, but you're probably not as mentally fit as you could have been. Bad air is not just a private problem for many children because they grow up in smoking households, it is now a problem of global proportions. According to a 2019 report from the US Health Effects Institute (which, interestingly enough, receives much of its money from the automobile industry), 95 percent of all children worldwide now breathe only polluted air.[23] According to the director general of the WHO, Dr. Tedros Adhanom Ghebreyesus, we are poisoning millions of children through exhaust fumes [actually there are over two billion] and ruining their lives.[24] They increasingly suffer from asthma and respiratory infections, which, according to WHO studies, killed over 600,000 children under the age of five in 2016 alone.[25] "Air pollution is one of the biggest threats to health of children in general," states the WHO. According to their investigation, one in ten deaths worldwide among children under the age of five is due to polluted air.[26] No matter how high the tax revenue from industry-friendly legislation may be, none of these children can be brought back to life.

According to the WHO report, the evidence is clear: "Polluted air has a devastating effect on the health of children." Air pollution, as it prevails especially in cities and metropolitan areas, has been shown to have harmful effects on pre- and postnatal brain development, according to studies by Amedeo D'Angiulli, a leading neuroscientist at Carleton University in Ottawa, Canada.[27] Among other things, it increases the likelihood of children developing ADHD or autism, both of which are frontal lobe maturation disorders.[28] There is also a direct relationship between air pollution and the need for psychiatric medication in children and adolescents.[29] In other words: *city life makes you mentally ill.* The burdening of children's lungs with particulate matter and nitrogen oxides on the way to school or when playing in the city park also demonstrably inhibits brain development and reduces academic success in the long term.[30]

Fine dust filter as a blood pressure reducer and dementia prophylaxis. Fine dust leads to many physiological regulation disorders and increases blood pressure. Filters against fine dust in the home caused by candlelight, cooking, or baking therefore reduce blood pressure significantly by up to 7.5 points.[31] In this way, physical and mental fitness could be improved in the long term without any side effects and could even protect against vascular dementia. In the long term, such filters would even prevent Alzheimer's disease, since fine dust also directly causes hippocampal functional impairment.[32]

Even a comparatively low level of air pollution is associated with a reduced volume of the frontal lobe.[33] In addition, increased exposure to particulate matter reduces hippocampal volume and memory performance.[34] Overall, increasing air pollution is thus a major reason for the global decline in average mental performance or intelligence, but also for the fact that bipolar disorder and severe depression are becoming more frequent. The increase in Alzheimer's disease is also due to increasing particulate matter pollution.

Escape from Reality

We damage our frontal lobe battery in many ways. We eat poorly, move too little, isolate ourselves and become lonely, increasingly lack meaningful goals, expose ourselves to enormous work-related and often social stress, and sleep poorly. In addition, we are chronically poisoning ourselves because we always give industrial interests priority over our health. At first glance, politics and business bear a large part of the responsibility here, but the list of supposedly voluntarily supplied toxins is also very long, so here are just two examples:

Alcohol acutely inhibits the release of the neurotransmitter glutamate at the hippocampal synapses and thus the ability to use System II.[35] This functional discharge of the frontal lobe battery can be quite intentional if one wants to escape the necessity of thinking about upcoming problems, which are usually emotionally upsetting for the person affected, and of making decisions. The alcohol-induced shutdown of System II also explains the old wisdom *in vino veritas*. It is incredibly difficult to be dishonest when under the influence of alcohol. Maintaining a facade requires a lot of mental energy, which an alcohol-impaired frontal lobe battery can no longer release (alcohol blocks hippocampal synapses). It reduces not only the ability to think, but also social inhibitions, both of which require System II.

A single instance of intoxication blocks hippocampal neurogenesis for several weeks,[36] which is why alcohol abuse particularly affects young people due to their very high rate of neurogenesis.[37] Adolescents with alcohol problems have very strong memory deficits and therefore often have problems at school; the same applies to university students. A group of US scientists summarized their results: "Our study shows that higher alcohol consumption during university studies is associated with a greater decline in GMV [gray matter volume, the number of brain cells] in the hippocampus, which in turn is associated with poorer memory performance, which ultimately has a significant impact on academic success."[38] In the long term, high alcohol consumption leads directly to Alzheimer's dementia,[39] whereby not only the frontal lobe battery is damaged but also the frontal lobe itself.[40]

Cannabis, or its active ingredient THC (tetrahydrocannabinol), leads to a loss of hippocampal volume and can lead to severe psychosis.[41] But it also causes many other brain changes, including thinning of the frontal cortex, which is a loss of nerve tissue in the executive center of our brain.[42] "This area of the brain is essentially involved in cognitive abilities such as impulse control, planning, problem-solving, prioritizing, and focusing," explains Dr. Maximilian Gahr, senior physician at the Clinic for Psychiatry and Psychotherapy III of the Ulm University Hospital, regarding a corresponding study.[43] The loss of these abilities is all the more pronounced the more cannabis is consumed—especially in adolescents. Since the development of the frontal lobe is not completed until very late, this is, according to Gahr, "probably the reason why the brains of adolescents react so particularly sensitively to drugs and other external disturbances."

But cannabis didn't always have such a fatal effect. Just a few decades ago, plants were grown with a largely balanced ratio of THC and cannabidiol (CBD), another active ingredient that has exactly the opposite effect: CBD promotes hippocampal neurogenesis and thus—examined in isolation—has anxiolytic and antidepressant effects.[44] CBD is therefore also a good candidate for supporting both systemic depression and Alzheimer's therapy.[45] However, consuming cannabis preventively is not advisable because for some years now, increasingly potent cannabis varieties with a very high THC content have dominated the market. These increase both the intoxicating effect and sales for drug manufacturers—but also the THC-related brain damage, especially as the protective CBD content was bred out.[46]

Chemistry in Food

Exposure to toxins makes sea fish, purely from a health perspective, inedible. This is due to high levels of methyl mercury and many other highly toxic industrial products that are carelessly dumped into the sea out of greed for money. For example, a mother's consumption of fish and seafood can have irreversible effects on the neurological development of her fetus, even if the mother does not experience symptoms.[47] But the products of modern agribusiness also have a lot to offer: Herbicides such as glyphosate and paraquat, as well as insecticides such as pyrethroids, organophosphates, neonicotinoids, and rotenone are some of their ingredients, and they all inhibit brain development and adult hippocampal neurogenesis.[48] It is therefore UGLY how politicians one-sidedly promote economic interests and continue to allow the use of these highly dangerous substances. This is how Dr. Johann G. Zaller of the Institute of Zoology at the University of Natural Resources and Life Sciences Vienna answered the question in 2017 as to whether there were still crucial gaps in knowledge regarding glyphosate that would justify delaying a ban: "In my opinion, supposed uncertainties in the assessment of the effects of glyphosate are deliberately spread by industry lobbies and exaggerated in the media. It is also remarkable with what patience politicians accept the trickery surrounding the approval of glyphosate that has been uncovered in recent months."[49] Since the limits are also adapted to the increasing use in agriculture (for example, the limit for glyphosate in soybeans was raised from 0.1 mg/kg to 20 mg/kg in 1999), the concentrations of glyphosate and AMPA (aminomethylphosphonic acid, the breakdown product of glyphosate, which is just as toxic) have been rising for decades in the urine of many people.[50] Zaller again says: "These findings show in a frightening way how contamination with glyphosate and its degradation product AMPA has increased in recent decades. Against the background of the proven health effects of glyphosate, of which the suspected cancer risks and nonalcoholic fatty liver disease are just the tip of the iceberg, one gets the impression that pesticides could be chronically poisoning us. The fact that the residues are below half the limit values only says that everything is within the legal framework. For substances such as glyphosate, which are also suspected of interfering with the hormonal system, these limit values are irrelevant because they work in very small quantities. In addition, the limit values only apply to individual substances, while we are confronted with hundreds of pesticides in everyday life." However, their use is still permitted—at least as of December 2022.[51]

But even if glyphosate is one day taken off the market, the conservative alternatives are no better. Zaller again: "I refuse to speculate on possible chemical alternatives to glyphosate because this automatically implies that pesticide-intensive agriculture is the only viable option. There are numerous non-chemical alternatives that have been used successfully for decades, for example in organic farming. This would obviously be a huge benefit for people and the environment. Due to the strong industrial and political promotion of conventional agriculture, research into non-chemical alternatives has been neglected for decades." I can therefore only strongly advise every reader to consume organic products as much as possible. Nevertheless, according to the Baden-Württemberg Organic Monitoring Report of 2012, organically grown products had 500 times lower pesticide contamination, and this low proportion has remained almost unchanged since then, as a further report from 2020 shows.[52] Two kilograms of conventional products contain as many toxins as one ton (a thousand kilograms) of organically grown food. The same amount of poison that you get in a week with conventional products takes around ten years with organic products. This makes it clear that responsibility for yourself and the next generations begins with nutrition. At the same time, you influence the entire environment with your purchasing decisions. Finally, each organic tomato grew in a small area of habitat that was neither over-fertilized nor contaminated with pesticides.

A healthy mind in an unhealthy environment is hardly possible. That's why it's dramatic for global brain health that the quality of drinking water is continually declining. I have already mentioned the high, brain-damaging nitrate pollution in the ground and drinking water caused by factory farming. But many harmful industrial products also reach our brains via the tap, such as the large chemical family of PFCs (perfluorinated compounds, i.e., perfluorinated substances), almost all of which are neurotoxic.[53] The reasons for the increase in the content of these pollutants, as well as for that of toxic heavy metals in drinking water, are manifold, but what they all have in common is our careless handling of nature.

A World Made of Plastic

Plastic residues in food also inhibit this growth of the brain precisely in the area that is responsible for cognitive functions at the highest level, such as error detection, conflict assessment, decision-making, and mental flexibility in general, as was found in animal experiments.[54] These results were confirmed in children.[55] The problem with so-called plasticizers (phthalates), from bisphenol A to Z to microplastics, starts very early. For example, baby bottles and plastic dolls are among the many abundant, toxic sources of these substances that even babies and toddlers come into contact with.[56] We are now even conditioned to perceive poison as the smell of something new and thus as something pleasant, as almost every buyer of a new car will admit. The new car smell is known to be highly toxic, and the air in the car exceeds the limits of many toxins that cause cancer for months.[57] But these also inhibit brain development, such as neurotoxic benzene[58] or formaldehyde,[59] which also passes through the placenta in pregnant women, easily reach the growing child's brain.

Basically, we should declare our immediate surroundings a plastic-free zone as much as possible. This may not be 100 percent successful, but less is more here. Plastic should never be heated. Plastic dishes should not be put in the dishwasher because this accelerates the release of the toxins they contain. To store food, cloth, paper, glass, or metal should be used as a health and environmentally friendly alternative to plastic.

Brain-Damaging Drugs

Many substances are prescribed to relieve symptoms of an unhealthy lifestyle, which do not make the basic problem disappear, but actually make it worse. Often others join in as "side effects." We have already talked about the brain-damaging effects of sedatives and sleeping pills from the benzodiazepine class (Valium, etc.). The *anticholinergics* form an even larger class. They are among the most frequently prescribed drugs worldwide, and they also damage the frontal lobe battery.

Anticholinergics are used, for example, to calm the stomach, bladder, or bronchi. In doing so, they inhibit the neurotransmitter acetylcholine, which is also an important messenger substance in the brain and a potent activator of hippocampal neurogenesis.[60] It is therefore to be expected that anticholinergic drugs have a detrimental effect on the development of the frontal lobe battery. In fact, a significant relationship is shown in retirement homes between the extent of intake (number as well as duration) of anticholinergics and the progression of mental decline.[61] According to the *Ärzteblatt* medical journal, this type of medication also includes tricyclic antidepressants, which, due to their anticholinergic properties, are used not only to treat depression, but also for anxiety and restlessness as well as sleep disorders.[62] Likewise, some first-generation antihistamines (also anticholinergic) are frequently administered because they are contained in over-the-counter sleeping pills (the most common active ingredients used in many sleeping pills are diphenhydramine or doxylamine). Furthermore, anticholinergics for cardiac arrhythmias, muscle tension, urinary urgency, and stomach ulcers, as well as frequently used antipsychotics or painkillers, contribute to the mental decline caused by medication. Listing all such substances is beyond the scope of this book—so if you are taking medications from this class, talk to your doctor about how to replace them.[63]

Acetaminophen (paracetamol, best known in the US as Tylenol) is the world's most widely prescribed drug for fever and mild pain. But it can also damage the brain in higher doses. And although paracetamol is considered a safe painkiller during pregnancy, it can inhibit frontal lobe development in the developing child and increase the rate of autism and ADHD.[64] The recommendation here is to take as little and as rarely as possible. In the long term, the intake leads to chronic inflammation,[65] especially in the area of the hippocampus.[66] But even a single intake reduces the ability to empathize, which is why the researchers involved in the study also refer to it as a "social analgesic [painkiller]": When under the influence of Paracetamol, we also perceive other people's pain less.[67]

Digital Detox

An experiment of enormous proportions is the increasing digitalization of our society. People are increasingly spending a large proportion of their time working on screens, which has a direct impact on brain function. Recently, a direct relationship was found between increased screen-based media use and reduced microstructural integrity of the white matter of the brain, the pathways through which different brain regions communicate internally with each other.[68] Fittingly, it was also found that this also impairs external communication, the language development of the preschool children examined, as well as their emerging reading and writing ability.

But the problem often starts much earlier. Parents today use digital media (television, computers, smartphones, tablets) for an average of nine hours per day, of which about three hours are spent on their smartphones.[69] Since parents—like all adults—are often distracted by the mobile media presence, studies show that they have fewer in-depth conversations, which leads to disruptions in the parent relationship and increased family stress.[70] It has even been shown that simply because screen time reduces direct eye contact, it also reduces the ability to empathize.[71] Thus, mobile phones inhibit relationship building because they reduce the individual's attention to their partner. It can even go so far that stressed parents calm their baby with a smartphone or tablet, which preprograms behavioral disorders. The two authors of a corresponding US study call the catastrophic long-term consequences "technoference," based on "interference," which refers to sustainably disturbed relationships, in this case through technology.[72] One of the consequences is a withdrawal of the children, but also more frustration, hyperactivity, whining, sulking, or tantrums can be seen. In addition, the little ones themselves are trained to use smartphones right from the start: "Children learn through imitation. When the toddler sees that my parents' most important item is their smartphone/tablet/PC, then the two-year-old is already asking for it. The Google-fication of socialization begins."[73]

Platforms such as Facebook, Instagram, WhatsApp, and most screen games are programmed for addiction and dependence, because they activate the reward system.[74] There is more dependency and an increase in over-the-top and aggressive behavior, but ultimately despite all the "social" activities that they supposedly partake in, numbers of very lonely children is increasing.

There is an alarming connection between the amount of time you use your smartphone and the likelihood of developing depression.[75] Attention and learning disorders are further consequences:

"Children become prisoners of the net and are at the mercy of the manipulations of IT corporations."[76] The WHO classifies *internet gaming disorder*, the addiction to constantly playing games on the internet, as an independent disease.[77] Every year, 20,000 children are newly diagnosed with media addiction. That is six times more new addicts every year than, for example, illegal drug consumption. "Media addiction is in some cases so manifest that the young people can no longer detach themselves from the media and can no longer pursue any education or work," reported congress director Dr. Uwe Büsching at the Adolescent Medicine Congress of the Professional Association of Pediatricians and Adolescent Physicians (BVKJ) in Weimar.[78] Well over 600,000 children were already media-addicted in Germany in 2017 according to this definition.

Permanent information overload is considered a major trigger for mental exhaustion and stress in the workplace. The ability to concentrate suffers, and the attention span continuously decreases.[79] All indications are that our frontal lobe is overwhelmed. "Digital detox" is primarily understood as an attempt to free yourself from the social pressure of constantly having to be present online and on social media. Based on the aphorism of Robert Browning (1812–1889):

> I give up the fight: If only there were an end,
> a seclusion, a dark corner for me.
> I want to be forgotten, even by God.

I would advise not to give up on the fight, but to reflect on your nature and natural needs. This includes taking time for yourself and reflecting and enjoying life without digitizing everything and making it available to the world online. However, smartphone addiction has become part of System I, a routine and felt basic need, so that it can only be stopped with a well-functioning System II, which most people now lack due to an increasingly low-capacity frontal lobe battery. In a *Spiegel* interview titled "This crap is corrupting us all!" (*Dieser Mist verdirbt uns alle!*), American software developer and internet critic Jaron Lanier explains the principle of "defend the beginnings": "All the parents who work at Google and Facebook don't allow their children to use the products they develop themselves. It's grotesque. The kids in Silicon Valley don't get cell phones and aren't allowed to sit in front of a screen. There are all these tech dads and tech moms, and they tell their kids: 'Careful, don't touch that, my company built it!' I think that does something to these parents. That really shakes them up."[80]

In my opinion, the digital detox should also include the radiation that comes from consumer devices and transmission towers. Animal experiments have shown that long-term exposure (i.e., 48 min/day for 30–180 days) to 900–1800 MHz from technology as early as the second generation of cell phone radiation, i.e. G2, damages the hippocampus.[81] G5 is at a higher frequency and therefore even higher energy reaching every corner of the earth. Even in your house you are no longer protected from radiation. Shearwood McClelland III and Jerry Jaboin from the department of radiation medicine at Oregon Health and Science University in their presentation on the expansion of global G5 and radiation safety implementation ask this question: Is it reassuring, or are we playing Russian roulette?[82] This question was also asked by Dr. JinHwa Moon of the University Children's Hospital in Seoul, Korea. She wrote about this in a comprehensive review: "The nervous system of children is more sensitive to the effects of electromagnetic waves (EMW) than that of adults. Although there are no studies on the effects of EMW on children's health, children's precautionary principles should be followed and children's exposure to EMW should be minimized. The fact that EMWs are potentially carcinogenic, according to the International Agency for Research on Cancer, should not be ignored or interpreted in a biased manner, and the opinions of clinicians should be given more weight than those of industry when establishing safety guidelines for EMW use. Studies are also needed to assess the effects of 5G frequency technology on children's health."[83] Among other things, long-term exposure to strong EMW in the home, school, or other places where children spend a large part of their time should be avoided. The UGLY is that governments are systematically ignoring these harmful effects when digitizing our society, as a US federal court found for the US—which will hopefully send a signal to other nations.

There are very many things that, if necessary, are directly within our personal power to change. For example, we don't have to wait for a law before we stop smoking in the presence of children. However, in order to protect them from particulate matter in cities that is at least as dangerous to health and brain-damaging, we need the power of society. And here the ultimate question arises: How does an increasingly frontal lobe-weakened society react to the increasingly urgent challenges that now threaten the very basis of life for all people?

CHAPTER 14

ZOMBIE APOCALYPSE— FINAL CHAPTER?

Yesterday I was still on the brink, but today I'm one step further.
—Unknown

Global Consequences of the Frontal Lobe Weakening

I'm sitting in our church, in the front row, on a hard, cold wooden bench. My first communion is coming up in a few months and I should believe in God. But the sacred stories that are supposed to help me do exactly the opposite. They seem rather unbelievable to me. The very fact that God sent his supposed son to the Middle East to save people from eternal damnation, while ignoring the rest of the world, sowed doubt in my mind. "God has given us free will," says the priest, "this is the only way we can freely decide for him." However, if we don't do it—he says in a preparatory Bible lesson—we are threatened with hell. Super. A huge problem for a child of just nine years old.

I still remember very clearly how I asked myself during the sermon: If I don't find the stories the priest tells us convincing and can't really believe them, is this inability just because I don't want to? Can I actually want what I want? These questions have bothered me since my youth. Perhaps Arthur Schopenhauer (1788–1860) became one of my favorite philosophers simply because he was the first to give me a logic that made sense to me and provide a convincing answer:

"Man can do what he wants, but he cannot will what he wants." Schopenhauer's answer to the question of whether we actually have free will was confirmed a good century later by neurobiological research: She also provided a clear "no" as an answer.[1] When I make a decision (such as believing in the biblical God or not), then I am aware of it, and in part, but only in part, I also know the reasons that moved me to decide in this way and not otherwise. But ultimately, my consciousness is only the user interface of a gigantic neural network that operates according to its own rules. Why I am aware of certain reasons but not others, how they were weighted in my decision-making process, these are all factors over which my ego has no control. You can't know what you don't know. The only thing that remains to me and for which my self is ultimately responsible is the final decision. I can do what I want, but I can't want what others think I have to want. If God had wanted us to actually have free will, he could have given us one, but only if he himself could really will, because he stands outside of nature.

Self-responsibility without free will. We are responsible for our actions, sometimes less, sometimes more. Less if we act in the heat of the moment (System I), under drugs, or in self-defense, more if we have had time to think with a clear mind (System II). From a scientific point of view, our will may not be free, but we still learn that all our decisions have consequences, for ourselves and for others. This creates responsibility.

However, we can do whatever we want. So at least we have freedom of action. However, this only exists if there are (a) several options for action and (b) at the same time, we also have the opportunity to weigh up and decide between them. But if we decide the same way again and again in similar situations because we lack the mental energy to think, then we ultimately even lack the ability to act freely. That is the dilemma of a person with a low-capacity frontal lobe battery, and thus of a large part of humanity. However, after the discovery of the frontal lobe battery and the neural correlate of its stored mental energy, it is evident that our ability to think rationally and make wise decisions is continuously diminishing due to our modern lifestyle. The frontal lobe battery cannot develop the capacity corresponding to its natural potential and then loses it even further, although it could increase it throughout life with a species-appropriate lifestyle, as Figures 25 and 26 illustrate in a "historical" comparison.

The more alien the way of life, the lower the frontal lobe battery capacity. As a result of the cultural development that can be observed, a large part of humanity is currently chronically exhausted. At the societal level, this leads to a shift to the left in the distribution of average frontal lobe strength, as Figure 26 shows. Shown graphically, like body weight and some other biological variables, frontal lobe strength is also distributed in a bell-like manner: Some people have a little more strength, others a little less. The majority typically fall somewhere in the middle between the extremes. If a large part of humanity changes their way of life from species-appropriate to species-unfriendly, as is the case today, a shift to the left inevitably occurs. We live in a society predominantly controlled by System I.

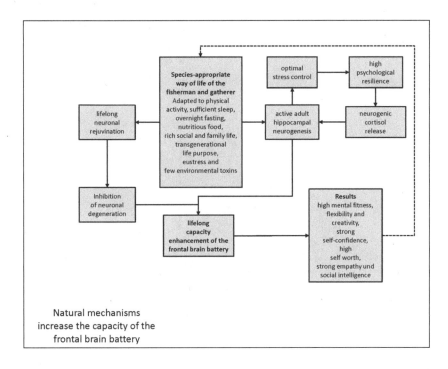

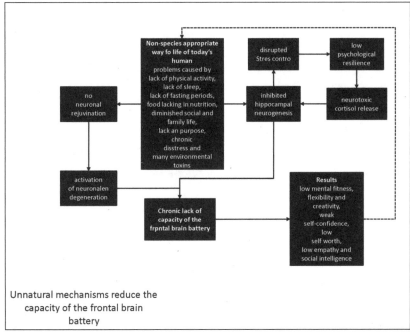

Figure 25: Frontal lobe battery capacity in species-appropriate and inappropriate/alien lifestyles.

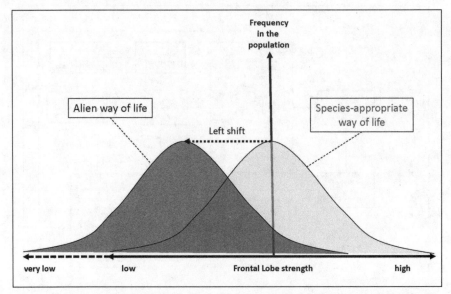

Figure 26: Schematic distribution of frontal lobe strength depending on today's lifestyle.

Of course, on an individual level, there are many reasons why people behave one way or another in one situation or another, just as many as there are individual fates. The psychological sciences and behavioral research have set themselves the task of qualifying, quantifying, and categorizing these. But although every person is unique and their behavior can usually be explained by their individual history, there are general tendencies in human behavior whose origins can be found in biological nature. This includes, for example, wanting to do everything necessary for one's children so that their future is secured. From this point of view, we would have to end the life-threatening air pollution and environmental destruction and reverse the process,[2] as well as the extinction of species and the contamination of the oceans. If this does not happen, if a large part of humanity is allowing the future of future generations to be destroyed with their eyes wide open, there could be a general explanation for this. Because despite all the individuality, in my opinion, much of what we observe in society as a whole follows a very simple logic. Brains, whose natural development is hindered by deficiencies in modern lifestyles and whose function is damaged by unfavorable environmental influences from which hardly anyone can completely escape, determine to a large extent the behavior of society as a whole. In particular, the prefrontal cortex and its batteries react particularly sensitively to defects or noxae (Latin *noxa* for damage, which includes any type of substance that is hazardous to health and harmful influence), which manifests itself in a weakening of their functions. Self-confidence and self-esteem decrease, which means those affected are more subject to pressure to conform to the group. There is also a general decline in social, emotional, and rational intelligence, as well as the ability to empathize. The IQ, as a measure of working memory located in the frontal lobe, is increasingly diminishing, as studies show.[3] Imagination, creativity, and fantasy suffer. Freedom of action is increasingly restricted. Humanity as a whole is becoming more and more incapable of breaking new ground or accepting and mastering new challenges when this is actually required, for its own good or for the good of others, especially future generations. Conservative thinking determines the social consensus.

Collectively, *Homo sapiens*, the wise man as we like to call ourselves, behave neither economically nor environmentally wisely. In a way, its behavior is reminiscent of that of the completely brainless yeast *Saccharomyces* (Greek-Latin "sugar fungus"): In a barrel filled with grape juice, this single-celled fungus multiplies exponentially by fermenting the glucose contained in it into carbon dioxide and alcohol to generate energy. This goes well until an alcohol concentration is reached that poisons him and ends his life and that of his offspring. He has thus erased himself. Although our "barrel" is larger and covers almost the entire surface of the earth, the motive for life of this single-celled organism does not differ from ours, not even in its ultimate consequence: It's about growth for everything in the world, until your own demise.

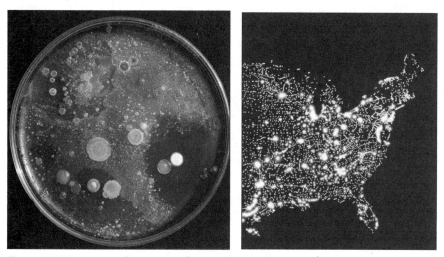

Figure 27: No smarter than a yeast fungus? Yeast colonies on a culture plate (left) and human colonies in North America (right; photographed at night from a satellite). Note the similarity of growth.

The Zombie in Us

Interestingly, people caught in routines don't stand out in everyday life, firstly because they belong to the vast majority and therefore represent what we consider normal, and on the other hand, because they can easily cope with their everyday lives using System I and are therefore on autopilot. As a reminder: System I also works without a frontal lobe battery. As long as everything goes in an orderly manner and it is enough to act in a stereotypical manner, everything is fine. All of our body cells and organs are constantly working without our mental input. It is only when something goes wrong, for example, that pain receptors draw our attention to a problem, that System II needs to address it. The same applies to the musculoskeletal system. The fact that you are currently tensing certain muscles to hold this book in your hands is probably only becoming clear to you now that I point it out to you. Sensors in muscles and tendons constantly send signals to the brain. But only when we consciously direct our inner feelers toward this neural activity do we feel, for example, how a tension develops in the neck. Then we consciously change the position. This happens automatically when you sleep. Here we change our position involuntarily. But even during the day, when we are awake, almost all of our mental activities occur unconsciously.

American neuroscientist Christof Koch and British physicist, molecular biologist, and Nobel laureate for the discovery of the nature of our genetic material Francis Crick (1916–2004), called this autopilot that controls our unconscious life the "zombie within us," in reference to the work of philosopher David Chalmers.[4] According to the *Stanford Encyclopedia of Philosophy*, "zombies are imaginary creatures that serve to illuminate problems of consciousness and its relation to the physical world."[5] Unlike zombies in movies, they are exactly like us in all physical aspects, but without consciousness. Nevertheless, their properties describe very well the majority of an existence controlled by System I.

The zombie in us masters all programmed, instinctive behavior patterns, such as sucking on the breast in early childhood. However, we have to consciously teach the zombie in us all the skills we need to learn. Only when he has mastered the technique and the processes have become "cortical reflexes" are we really good at something. However, since these reflexes are not conscious, we usually do not remember how we did something with the help of our zombie. We see the result of the action and remember it if it was significant, but not the actual action. This applies to playing tennis and driving as well as to reading and writing. To be able to do this, the zombie within us receives information from all our sensory organs and uses it to control our muscles completely unconsciously. In this way, he ensures that we do not collapse when we think of other things during routine work or simply daydream. In many cases, the zombie in us is far superior to our conscious actions. For example, fill a glass full to the brim with water and walk around your apartment with it. You'll most likely spill less water if you don't focus on it. The zombie in us can do better. Or suppose you play tennis. To your chagrin, your opponent plays an almost flawless backhand. Congratulate him on this and emphasize his graceful swing. It could well be that from then on he will have difficulty playing a good ball. Because of the praise, he now consciously perceives his backhand and therefore loses his "flow." They sent his zombie to the bench.

For Koch and Crick, "the hallmarks of the zombie system are its stereotyped, limited sensorimotor behavior [this means that sensory input and motor output are directly linked as cortical reflexes] and thus immediate, rapid action." However, its generally excellent functioning in everyday life raises two central questions for the two scientists: "Why aren't we just big bundles of unconscious zombies? Why has consciousness evolved that takes hundreds of milliseconds to get going? Perhaps," the two scientists answered, "because consciousness allows us to plan future actions, which opens up a potentially infinite behavioral repertoire."[6]

According to the latest theory, conscious memories and ultimately consciousness actually only arise when the predictions of System I fail and the zombie within us encounters unforeseen problems.[7] So whenever we can no longer continue according to old habits, or should not, without possible threat of danger. We live largely unconsciously as long as we don't switch on System II, which is why we usually don't remember it when we go about our everyday routines. We are often unsure whether we actually turn off the stove after cooking or lock the front door before going to bed. There is no explicit reminder; we could have been sleepwalking during this time. Because we go through life mostly unconsciously, our brain prevents the mental battery from discharging. A conscious life is a stressful life, and our brain wants to avoid that in order to save mental energy. This is how the South African brain researcher and neuropsychoanalyst Mark Solms explains it: "The brain would prefer nothing unexpected to happen. Total uniformity is much more conducive to life than consciousness [conscious reflection using System II], which robs energy and time."[8] Ideally, however, a gut feeling, System I, signals to us that we should actually think a little longer and more intensively about solving a pending problem using System II. Part of this gut feeling—at least that's what I noticed for myself—is the brief switching on of System II when switching off the stove and locking the door in the evening. By becoming aware of these processes or "switching on" my consciousness, I create hippocampal memories of the action and in this way prevent having to get out of bed again to check the situation (which also requires energy, albeit a purely metabolic one). Our consciousness therefore needs a quick memory, the ability to remember our own actions or thoughts. Conversely, this means that with the chronic loss of frontal lobe battery capacity, which is part of our normality, a large part of humanity is increasingly acting only zombielike. It is then only a small step from individual self-destruction, the loss of individuality through chronic depletion of the frontal lobe battery, to global self-destruction. The loss of empathy associated with the shrinking of social memory exacerbates this fatal development, as we then become more and more indifferent to the consequences of our own actions for others.

Loss of Battery Capacity—Loss of Empathy

If you look at the currently prevailing behavior in our modern society, it is characterized by ignorance toward the vital challenges. Too often the ability to learn from past mistakes, develop new concepts, and bring about necessary changes as efficiently as necessary is missing. What is also noticeable is the increasing social coldness.

People who lack empathy lack the basic ability to imagine the potential suffering their actions may cause. When the ability to empathize decreases, suffering increases. To measure and quantify empathy, Mark H. Davis, professor of psychology at Eckerd College in Florida, developed the Interpersonal Reactivity Index (IRI).[9] The IRI allows a multidimensional assessment of a person's capacity for empathy. In a comprehensive analysis encompassing a total of seventy-two IRI studies, the development of empathy was determined in US college students—representative of an industrialized society of the global north—over a period of thirty years (1979 to 2009).[10] Overall, the authors of the meta-study found a significant decrease in the students' ability to empathize with other people. The most noticeable reduction, however, was the ability to feel compassion for other people who were less fortunate than oneself.

Interestingly, some scientists are pointing to an increasing lack of empathy in human beings in society as being also responsible for the destruction of the biosphere and biodiversity.[11] Gerardo Ceballos of the National Autonomous University of Mexico writes that "the massive loss of populations and species reflects our lack of compassion for all the wild species that have been our companions since our emergence."[12] According to Ceballos, this is all just a prelude to the "decline of the natural systems that make our civilization possible." The extent of the global destruction of species is now being compared to the fate that befell not only the dinosaurs but over 90 percent of all life 65 million years ago—only this time the catastrophe will be caused not by a huge meteor, but by humans.

A lack of empathy is also typically associated with an increased tendency toward narcissism. This is characterized by an excessive focus on the ego and thus an exaggerated tendency to put one's own well-being above that of others. In fact, there was a measurable increase in narcissistic tendencies in almost the same period as the IRI studies.[13]

Weak I Seeks Strong We

Our ego, what we identify with, is the entirety of our individual, personal memories—the conscious ones, but also the hidden ones. As the hippocampus shrinks, our ego or elementary parts of ourselves also shrink. In a sense, we die a small death every day. The loss of hippocampal individuality is accompanied by a loss of ego strength and self-esteem. This development activates a survival instinct: The weak I looks for a strong We. People with low self-esteem tend to associate with others in order to protect their weakened self in an often exuberant We. The result is often what is known as collective narcissism. One's own group, with its mostly group-specific stereotypes through which its members ultimately define themselves, is perceived as unique. It wouldn't be a bad thing if it stayed that way. But unfortunately, narcissistic communities not only bathe in their own glory, as psychologist and science journalist Corinna Hartmann explains, they also hold a grudge against anyone else who does not recognize this uniqueness.[14] Those who think differently are labeled outsiders, isolated and less and less seen as part of society.

"In contrast to individual narcissists," says Hartmann, "collective narcissists derive their self-worth more from their social self: the identity that membership in the supposedly superior group gives them."

By creating images of the enemy, you strengthen your own group and thereby create compliant behavior. Even something as absurd as xenophobia is legitimized by the respective group: if the *We* agrees, the *I* cannot be wrong. It should be noted here, however, that frontal lobe weakness per se does not automatically lead toward a collective narcissism—and certainly not toward a supposedly legitimized xenophobia or conservative nationalism. But people with reduced mental performance, will, and above all ego strength are easier to manipulate because there is hardly any questioning; there is a lack of foresight and healthy self-esteem.

The pandemic of self-weakening and the closely related global increase in collective narcissism thus increase the likelihood that the world community will ultimately fail to improve the living conditions of all people and to stop the ongoing global environmental destruction and many other serious problems that endanger the future of our children. Since no nation is solely responsible for environmentally destructive developments, no one can stop them on their own. But perhaps for many people the prospect that there will soon be artificial intelligence that can solve all of our problems feels like a stroke of luck.

From Idiocracy to Infocracy

Anyone who doesn't think for themselves will be thought for—this also applies to politics. Back in 2010, then Google boss Eric Schmidt made it clear in an interview at the Washington Ideas Forum: "The laws are written by lobbyists."[15] According to Schmidt, one thing is certain: "the average American [and certainly also the average German] is not aware of how much of the law is written by lobbyists to protect the interests of established companies." Visibly enthusiastic, he added, "It's shocking how the system actually works." Google, Facebook, Amazon, etc., collect data about everything and everyone and are therefore the world's most influential companies. Knowledge is their power and data about us is their powerful arsenal.

Even then, Schmidt envisioned a future in which machines and technology would play the decisive role in our lives: "With your permission, you will give us more information about you and your friends and we can improve the quality of our searches," he said. "We will no longer need you to enter data. We know where you are. We know where you've been. We can more or less know what you are thinking of." But when machines actually know what we think and what we want (maybe even before we know it ourselves), then everything changes. Then it is only a small step until they completely take over the thinking for us, until we are completely incapacitated. Eric Schmidt: "I think most people want . . . Google to tell them what to do next." Peter Hensinger, educator and board member of Diagnose-Funk eV, says in his article "The Digital Pandemic": "Always knowing where every person is, what they are doing, and what their biological state is, is the DNA of the new surveillance capitalism that no longer knows any choices."[16] Indeed, according to Schmidt's vision of the future, technology will also "completely change the way government works." This has now also found its way into the planning of cities; at least there are concrete visions. The key term is "smart city." First of all, it's about developing smart cities that will ultimately be smarter than its residents.

The year is 2017. Under the "Smart City Charta" concept, the Federal Ministry of the Interior and Community (BMI) presents how the "digital transformation in municipalities could be sustainably designed." The basis for this is a keynote speech by Roope Mokka, founder of the Finnish think tank Demos.[17] The term "charta" was perhaps not chosen by chance, as it means something like "legally binding basic order." Here within, the chapter "Internet of NO things" states: "On average, we look at our smartphones every six minutes: We pause what we're doing every six minutes for Facebook, Snapchat, Instagram, Twitter, and Reddit. This is, of course, crazy—but maybe not as crazy as saying that in ten years we won't be looking at our smartphones at all. However, this is exactly what will happen. This is because smartphones will no longer exist: Their functions will be integrated into our environment—in just ten years."[18] This would mean Eric Schmidt's vision of complete digital surveillance has been achieved, and privacy would be a thing of the past. Everything we do, think, and feel would be recorded and exploited. That's not a by-product, it's inevitable. In the chapter "Visions of a Hyper-Connected Planet," Mokka describes the dangers that the "Internet of NO things" can bring.[19] Here are, in my opinion, the three most relevant of the six points to be read there:

- *Post-choice society*—Artificial intelligence replaces our freedom of action, but expressed positively. "We never have to decide whether to take a particular bus or train, but are shown the quickest route from A to B. We will never forget our keys, wallets, or watches."
- *Post-ownership society*—Thanks to the information about available shared goods and resources, it makes less sense to own something. "Perhaps private property will indeed become a luxury. Data could complement or replace money as a currency."

- *Post-voting society*—"Since we know exactly what people do and want, there is less need for elections, majority determinations, or votes. Behavioral data can replace democracy as society's feedback system."

If what Roope Mokka describes here becomes a reality, we will essentially be living in a zoo, with an artificial intelligence (AI) as the keeper. It determines who gets what to eat and who can live where. Whether you can take a train ride to visit friends, family, or on vacation (from what?). In order for this to happen fairly, a social scoring system like the one we know from China would be entirely conceivable: Only the good monkey gets the banana. "Behavioral data" would be used. About a year after the publication of the Smart City Charta, *Stern* editor Gernot Kramper reported on the preparatory work toward a "smart society" that China has already done: "Good behavior should be rewarded, bad behavior should not. The Chinese Communist Party wants to introduce this simple education law nationwide [this happened in 2020]: All citizens should be evaluated. The 'Social Credit System' is a kind of credit score for all aspects of life." According to Kramper, this has never been done before. "No country has come as close to the vision of a comprehensive educational dictatorship as China."[20] To this end, "given China's impartiality when it comes to data protection . . . all data that is available to the state will be included in the assessment of the population. People who don't pay their bills or are prone to bar fights are likely to quickly be downgraded to second-class citizens." In Rongcheng, a large city in the northeast of the country, the system has been tested on a large scale since 2014. "There, 'bad' citizens," says Kramper, "can no longer buy tickets for achievements such as high-speed trains or plane tickets."

Chinese President Xi Jinping expressed himself quite positively about this total digital surveillance and evaluation: "A feeling of security is the best gift that a country can give its citizens."[21] Is such a development only possible in countries like China? Not necessarily. Anyone who knows the past well can also think about future developments in a frighteningly plausible way. Thus, according to historian Yuval Noah Harari, the COVID crisis is indeed "turning the world upside down, but it may only mark the beginning of a new era: that of total surveillance of all people."[22] Although, historically speaking, this pandemic is not as dangerous as the plagues of the past, the political and economic impact of the COVID pandemic, according to Harari, could be enormous: "In the worst case, our world order collapses." In his opinion, in fifty years, "people will not remember the epidemic itself so much. Instead, they will say: This was the moment when the digital revolution became reality . . ., that in the year 2020, with the help of digitalization, the ubiquitous surveillance by the state began." Not only in China, but also in the West, this would be quite conceivable—the coronavirus also acts globally. Harari says: "I'm not sure this scenario will happen. But I fear that total control could be a consequence of the Corona crisis. Many things that were unthinkable in the West just a year ago. The pandemic has suddenly made them acceptable there too."

With social scoring likely to become a reality for us in the future, as an internet user you should know that anything you do there can and probably will be used against you. However, this will not be used against us in a regular court (an AI can do without a judge and jury), but in determining our future living conditions. AIs record everything (everything future, but also everything past), and everything has an impact. Which apartment we are allowed to live in, which job we or our children are allowed to do, whether we will be looked after if we still get sick despite (forced) vaccinations against everything that could make us sick. Maybe the AI will even speak to us in the form of a chatbot that can take on any shape.

"The containment of the corona pandemic requires a global surveillance network," state Klaus Schwab, German economist, founder and executive chairman of the World Economic Forum, and Thierry Malleret, French economist and cofounder and lead author of the *Monthly Barometer*, an analytical and prognostic newsletter on macro issues for high-level decision-makers, in their book *COVID-19: The Great Transformation*, which is very reminiscent of the Smart City Charta in many respects.[23] A government study by the BMI (Federal Ministry of the Interior) from the beginning of the pandemic in March 2020 said similarly: "In order to make testing faster and more efficient, the use of big data and location tracking is essential in the long term."[24]

In practical terms, twenty-four-hour digital surveillance is no longer a problem these days, especially if everyone always carries their spy device in the form of a smartphone with them. "The evaluation of the data volumes can," says Harari, "be taken over by an artificial intelligence that can even calculate how a monitored person will likely behave in the future." Quite in the spirit of former Google CEO Schmidt, Harari then also predicts: "For the first time in history, total surveillance is possible. You can learn more about people than they know about themselves. This is the real danger that the current crisis brings with it: That digital surveillance technology through the health crisis legitimized worldwide—even in democratic societies that previously resisted surveillance."

Media control: How the media manipulates us. Noam Chomsky, one of the world's best-known intellectuals, writes in his book of the same name: "Citizens of democratic societies should attend courses in mental self-defense to be able to defend themselves against manipulation and control."[25] According to Chomsky's analysis, the media also serve, without being subject to direct state control, "the social production of consensus, suppressing news that could unsettle the population, mitigating it so that there is no doubt about the attitude of the political leadership."

The enemy image of COVID allows the governments of the world to advance digitization toward the completely transparent human without encountering too much resistance—not thinking is the motto of the times, we only act with System I. I wonder, among other things, what will happen if future AIs, which could control our lives, begin to "think" using System II. In fact, scientists believe that consciousness, i.e. self-reflection, could also be possible for intelligent machines. At least this is being seriously researched.[26]

One of the most prominent researchers is Ray Kurzweil, considered by many to be the world's leading AI visionary. He has been working at Google since the end of 2012 on the company's next big goal, namely "an artificially intelligent search engine that knows us better than we know ourselves."[27] In his opinion, and he has rarely been wrong in his future predictions, computers will be smarter than humans by 2029 and will make jokes and flirt. Nobody knows whether they will laugh with us or at us. Whether they will still need us then is equally irrelevant.

I think, therefore I am. *But am I still*, the French philosopher, mathematician, and scientist René Descartes (1596–1650) would have to ask himself today, *if I leave the thinking to an artificial intelligence?* Whether the apocalypse of AI-monitored, zombie-like life will be the final chapter of human history is difficult to say.

"If it doesn't succeed," you read in the journalistically independent magazine *OPEN Insights*, "If you write human values into the core code, so to speak, into the 'DNA' of a super-intelligent computer from the start, the AI will almost inevitably come to the conclusion that people are standing in its way at some point. And that could mean the end of humanity."[28] Then artificial intelligence would actually be our last invention. Theoretical physicist and astrophysicist Stephen Hawking (1942–2018) said something similar in the same year when the BMI published the Smart City Charta: "You can imagine that such a technology [AI] outsmarts the financial markets, outsmarts human researchers, manipulates human leaders, and develops weapons we don't even understand. While the short-term impact of AI [development] depends on who controls it, the long-term impact depends on whether it can be controlled at all."[29]

Albert Einstein (1879–1955) is credited with having a premonition: "I fear the day when technology surpasses our humanity. There will only be one generation of idiots left in the world." Now that I have written this book, I have come to the conclusion that this vision of the future can be interpreted in a different way and somewhat less favorably could be formulated: "A generation of idiots allows technology to surpass our humanity." Given recent cultural developments, Einstein might agree with me.

Hope Dies Last

When I published my first book on evolutionary disease prevention based on evolutionary biology in 2011, my hope was that if humans found their way back to nature, they would not only heal themselves,[30] they would also protect nature and preserve the life opportunities of future generations.[31] This was a bit naive, which I was well aware of, because the book begins with the words: "There is a mass extinction taking place in Germany and no one cares . . ." Obviously, only short-term thinking and acting usually determine our actions. Diseases of civilization are not avoided by changing your lifestyle and thereby eliminating the causes, but rather treated by alleviating the symptoms. Unfortunately, there was no willingness to make urgently needed changes on a broad social level to the extent hoped. Even the widespread fear of Alzheimer's became a motivator for only a few to seek protection through a more natural way of life.[32] My last hope was the COVID crisis—as with Alzheimer's, there is no effective medicine against the changing virus. It is very likely that the next mutation that can circumvent possible vaccination protection is already on the global path. Almost eight billion people are impossible to protect with the measures taken so far. Since herd immunity is not achievable, but herd health is, the pandemic seemed to me to be a welcome opportunity to address the health weaknesses of our society that the virus exposes.[33]

But unfortunately, there is little interest in eliminating the true causes of the pandemic. The virus thus exposes a completely different weakness in our society: the loss of human wisdom. Instead of *Homo sapiens*, the wise person, we only seem to be *Homo insipiens*, unable to learn and lacking insight. At school I learned a Latin proverb called *sapereaude*, which means "dare to be wise." The great Enlightenment thinker Immanuel Kant (1724–1804) made it, slightly modified, the motto of the Enlightenment: "Have the courage to use your own understanding!" In Kant's time there were probably still enough people who had the courage to do so and stood up for more freedom and civil rights, which we (still) have today. In his own words, the age of enlightenment that he ushered in with his writings was "the emergence of man from his self-inflicted immaturity. Immaturity is the inability to use one's understanding without the guidance of another." But with dwindling frontal lobe batteries, independent thinking becomes an effort, and fear reigns. Little by little we are allowing our civil rights to be taken away and we are increasingly slipping into complete immaturity. In this way, we betray the spirit of the Enlightenment, whose sociopolitical effects and freedoms we were able to enjoy for a long time. Instead of defending ourselves, as Kant would have demanded, we follow the frightened flock like sheep.

The correct measures against COVID, correcting the micronutrient deficiencies responsible for the severe courses of infection, which would lead to "coronary" herd immunity, would also be the first correct measures to remedy the frontal lobe weakness of society. We missed this opportunity, but it is not too late. I'm an eternal optimist. Perhaps there is still a chance, but this hope is only somewhat clouded by the fact that the measures that were taken (such as social isolation and lockdowns with the consequence of more lack of exercise and malnutrition as well as frightening the population) have most likely driven the general frontal lobe weakness. A huge increase in mental illness is either already being seen or is expected in the near future.[34]

But we actually have no alternative. All other global challenges, such as sustainably securing species-appropriate nutrition for the world, ending progressive environmental destruction, and introducing social justice and opportunities for the future for *all* people, also require meaningful local and global solutions. Mastering these will only succeed with a functioning socially and sustainably thinking and acting frontal lobe.[35] This is also the conclusion reached by empathy researcher Helen Riess in her article on the "Science of Empathy": "If we want to move towards a more empathic society and a more compassionate world, it is of crucial importance for strengthening individual, communal, national, and international bonds to strengthen our empathic abilities." A different way of thinking would be urgently needed.

Do we have a chance? Yes, we always have a chance, because as we all know, hope dies last. Is it realistic? That depends on how many of us want to take that chance. I recently read an insight from an unknown author that seems relevant to me here:

"That's not possible," everyone said.
"Then someone came along who didn't know that and just did it."

IN GRATITUDE

This book is based on a vast collection of scientific data that countless researchers around the world have compiled and published. Therefore, as with my previous works, my first thanks go to the generations of scientists who have created this valuable foundation. I received many helpful comments and suggestions from Bettina Simonis and my wife Sabine, both of whom supported me greatly during the writing process. Special thanks also go once again to my editor at Heyne, Friederike Achter, and to the editor Sophie Dahmen, who, through wonderful teamwork, made the book so much better.

RECOMMENDED READING

Aufschnaiter, U. *Deutschlands Kranke Kinder. Wie auf Anweisung der Regierung Kitas und Schulen die Gesundheit unserer Kinder schädigen*, Tredition: Hamburg, 2019.

Bostrom, N. *Superintelligence: Paths, Dangers, Strategies*, Oxford University Press: Oxford, 2014.

Böttcher, S. *Wer, wenn nicht Bill? Anleitung für unser Endspiel um die Zukunft*, Rubikon: Mainz, 2021.

Brown, S. and Vaughan, C. *Play: How It Shapes the Brain, Opens the Imagination, and Invigorates the Soul*, Avery: New York, 2010.

Busse, T. *Die Wegwerfkuh. Wie unsere Landwirtschaft Tiere verheizt, Bauern ruiniert, Ressourcen verschwendet und was wir dagegen tun können*, Blessing: Munich, 2015.

Chomsky, N. *Media Control: The Spectacular Achievements of Propaganda*, Seven Stories Press: New York, 2002.

Chomsky, N. *Who Rules the World?* Penguin: Munich, 2017.

Cortright, B. *Das bessere Gehirn. Wie Sie lebenslang die Bildung neuer Hirnzellen anregen*, Scorpio: Munich, 2017.

Crouch, C. *Post-Democracy*, Polity: Cambridge, 2004.

von Ditfurth, C. *Wachstumswahn. Wie wir uns selbst vernichten*, Lamuv: Göttingen, 1995.

Engel, V. *Am Anfang war die Mutter—Persönlichkeitsbildung. Ein allgemein verständlicher Leitfaden aus naturwissenschaftlichtiefenpsychologischer Sicht*, V. Engel: Konstanz, 2021.

Frenkel, B. and Randerarth, A. *Die Kinderkrankmacher. Zwischen Leistungsdruck und Perfektion—Das Geschäft mit unseren Kindern*, Herder: Freiburg, 2015.

Hartmann, K. *Aus kontrolliertem Raubbau. Wie Politik und Wirtschaft das Klima anheizen, Natur vernichten und Armut produzieren*, Blessing: München, 2015.

Hartmann, K. *Die grüne Lüge. Weltrettung als profitables Geschäftsmodell*, Blessing: München, 2018.

Holt, J. *Gibt es alles oder nichts? Eine philosophische Detektivgeschichte*, Rowohlt: Reinbek, 2014.

Kahneman, D. *Schnelles Denken, langsames Denken*, Siedler: München, 2012.

Kollektiv ILA (Hg.). *Auf Kosten Anderer? Wie die imperiale Lebensweise ein gutes Leben für alle verhindert*, Oekom: München, 2017.

Kreiß, C. *Gekaufte Forschung. Wissenschaft im Dienst der Konzerne*, EuropaVerlag: Berlin, 2015.

Mau, S. *Das metrische Wir. Über die Quantifizierung des Sozialen*, Suhrkamp: Berlin, 2017.

Mausfeld, R. *Warum schweigen die Lämmer? Wie Elitendemokratie und Neoliberalismus unsere Gesellschaft und unsere Lebensgrundlagen zerstören*, Westend: Frankfurt, 2018.

Meadows, D. *Die Grenzen des Wachstums. Bericht des Club of Rome zur Lage der Menschheit*, DVA: München, 1972.

Nehls, M. *The Algae Oil Revolution*, Skyhorse: New York, 2025.

Nehls, M. *Die AlzheimerLüge. Die Wahrheit über eine vermeidbare Krankheit*, Hey ne: Munchen, 2014.

Nehls, M. *Das CoronaSyndrom. Wie das Virus unsere Schwächen offenlegt—und wie wir uns nachhaltig schützen können*, Heyne: Munchen, 2021.

Nehls, M. *Herausforderung RACE ACROSS AMERICA. 4800 km Zeitfahren von Küste zu Küste*, Mental Enterprises: Vorstetten, 2012.

Nehls, M. *Die MethusalemStrategie. Vermeiden, was uns daran hindert, gesund älter und weiser zu werden*, Mental Enterprises: Vorstetten, 2011.

Pellis, S. and Pellis, V. *The Playful Brain: Venturing to the Limits of Neuroscience*, One World Publications: London, 2010.

Pinker, S. *The Blank Slate: The Modern Denial of Human Nature*, Viking: New York, 2002.

RenzPolster, H. *Kinder verstehen. Born to be wild. Wie die Evolution unsere Kinder prägt*, Kösel: München, 2009.

Roth, J. *Der stille Putsch. Wie eine geheime Elite aus Wirtschaft und Politik sich Europa und unser Land unter den Nagel reißt*, Heyne: München, 2016.

Sapolsky, R. *Gewalt und Mitgefühl. Die Biologie des menschlichen Verhaltens*, Carl Hanser: München, 2017.

Schmidt, N. and Meitert, C. *artgerecht—Das andere BabyBuch. Natürliche Bedürfnis se stillen. Gesunde Entwicklung fördern. Naturnah erziehen*, Kösel: München, 2015.

Schmidt-Salomon, M. *Manifest des evolutionären Humanismus. Plädoyer für eine zeit gemäße Leitkultur*, Alibri: Aschaffenburg, 2006.

Solms, M. *The Hidden Spring: A Journey to the Source of Consciousness*, Profile Books: London, 2021.

Voland, E. *Die Natur des Menschen. Grundkurs Soziobiologie*, C.H. Beck: München, 2007.

Wallert, M. *Strenth Through Crises: The Art of not Losing Your Head*, Ullstein, 2021.

Wilson, E. O. *The Social Conquest of the Earth*, C.H. Beck: München, 2013.

Witzer, B: *Die Diktatur der Dummen. Wie unsere Gesellschaft verblödet, weil die Klügeren immer nachgeben*, Heyne: München, 2014.

NOTES

INTRODUCTION

1   Popper, KR. Alles Leben ist Problemlösen: Über Erkenntnis, Geschichte und Politik, Piper: München, 1996.

CHAPTER 1

1   Gleick, J. *Chaos: Making a New Science*, Penguin Books: Melbourne, 1987.

2   Runco, MA, and Jaeger, GJ. "The standard definition of creativity," *Creativity Research Journal* 2012, 24:92–96.

3   Viereck, GS. "What Life Means to Einstein," *The Saturday Evening Post*, October 26, 1929.

4   Damasio H. et al. "The Return of Phineas Gage: Clues about the brain from the skull of a famous patient," *Science* 1994, 264:1102–1105.

5   Van Horn, JD et al. "Mapping connectivity damage in the case of Phineas Gage," *PLoS One* 2012, www.ncbi.nlm.nih.gov/pubmed/22616011.

6   Firat, RB. "Opening the 'Black Box': Functions of the Frontal Lobes and Their Implications for Sociology," *Frontiers in Sociology* 2019, http://www.frontiersin.org/artic les/10.3389/fsoc.2019.00003/full.

CHAPTER 2

1   Baumeister, RF et al. "Ego depletion: Is the active self a limited resource?" *Journal of Personality and Social Psychology* 1998, 74:1252–1265.

2   Kahneman, D. Schnelles Denken, langsames Denken, München: Siedler, 2012.

3   Danziger S et al. "Extraneous factors in judicial decisions," PNAS USA 2011, 108: 68896892.

4   Glöckner, A. "The irrational hungry judge effect revisited: Simulations Reveal that the magnitude of the effect is overestimated," *Judgment and Decision Making* 2016, 11: 601–610.

5   Ewen, E. and Ewen, S. Typen und Stereotypen: Die Geschichte des Vorurteils, Berlin: Parthas, 2009.

6   Le Pelley, ME et al. "Stereotype formation: biased by association," *Journal of Experimental Psychology: General* 2010, 139:138–161.

7   Muraven M. "Prejudice as Self-control Failure," *Journal of Applied Social Psychology* 2008, 38:314333.

8   Kemp R et al. "Willpower building: A new element in relapse prevention," *Health Psychology Report* 2016, 4: 281–293.

CHAPTER 3

1   Sperber, D. "What the judge ate for breakfast," International Cognition and Culture Institute, April 13, 2011, https://cognitionandculture.net/blogs/dan-sperber/what-the -judge-ate-for-breakfast/index.html.

2   Madsen, P. L. et al. "Persistent resetting of the cerebral oxygen/glucose uptake ratio by brain activation: Evidence obtained with the Kety-Schmidt technique," *Journal of Cerebral Blood Flow and Metabolism* 1995, 15: 485–491.

3   Kurzban R. "Does the brain consume additional glucose during self-control tasks?" *Evolutionary Psychology* 2010, 8:244259.

4   Tomporowski, PD. "Effects of acute bouts of exercises on cognition," *Acta Psychologica* 2003, 112:297–324; Hillman, CH et al. "Be smart, exercise your heart: Exercise effects on brain and cognition," *Nature Reviews Neuroscience* 2008, 9:58–65.

5   Hillman, CH et al. "The effect of acute treadmill walking on cognitive control and academic achievement in preadolescent children," *Neuroscience* 2009, 159:1044–1054.

6   Lange F et al. "Turn It All You Want: Still No Effect of Sugar Consumption on Ego Depletion," *Journal of European Psychology Students* 2014, 5:18; Lange F and Eggert F. "Sweet Delusion: Glucose drinks fail to counteract ego depletion," Appetite 2014, 75: 5463.

7   Beilharz, JE et al. "Short-term exposure to a diet high in fat and sugar, or liquid sugar, selectively impairs hippocampal-dependent memory, with differential impacts on inflammation," *Behav Brain Res* 2016, 306:1–7; Kerti, L. et al. "Higher glucose levels associated with lower memory and reduced hippocampal microstructure," *Neurology* 2013, 81:1746–1752.

8   Dextro Energy School Material Forest Fruit, World of Sweets, https://www.world ofsweets.de/DextroEnergySchulstoff Wald frucht50g.307170.html.

9   Vadillo MA et al. "The Bitter Truth About Sugar and Willpower: The Limited Evidential Value of the Glucose Model of Ego Depletion," *Psychol Science* 2016, 27:1207–1214.

10   Lashley, KS. "Physiological mechanisms in animal behavior," *Society for Experimental Biology* 1950, 4:454–482.

11   Scoville, WB and Milner, B. "Loss of recent memory after bilateral hippocampal lesions," *Journal of Neurology, Neurosurgery and Psychiatry* 1957, 296:1–22.

12   Corkin S. "Lasting consequences of bilateral medial temporal lobectomy: Clinical course and experimental findings in H.M.," *Sem Neurol* 1984, 4:249259.

13   Hilts, PJ. *Memory's Ghost: The Nature Of Memory And The Strange Tale Of Mr. M.*, Touchstone: New York, 1996, 111.

14 Hoffmann, M. "The human frontal lobes and frontal network systems: an evolutionary, clinical, and treatment perspective," *ISRN Neurology* 2013, www.ncbi.nlm. nih.gov /pubmed/23577266.

15 Triarhou, LC. "The signalling contributions of Constantin von Economo to basic clinical and evolutionary neuroscience," *Brain Research Bulletin* 2006, 69:223–243.

16 Allman, JM et al. "The Von Economo neurons in the fronto-insular and anterior cingulate cortex," *Annals of the New York Academy of Sciences* 2011, 1225:59–71.

17 González-Acosta CA et al. "By Economo Neurons in the Human Medial Frontopolar Cortex," *Front Neuroanatomy* 2018, www.ncbi.nlm.nih.gov/pmc/articles/PMC6087737.

18 Oehrn CR et al. "Human Hippocampal Dynamics during Response Conflict," *Current Biology* 2015, 25:23072313.

19 Santillo, AF et al. "Von Economo neurons are selectively targeted in frontotemporal dementia," *Neuropathology and Applied Neurobiology* 2013, 39:572–579.

20 "The Nobel Prize in Physiology or Medicine 1949," www.nobelprize.org/prizes /medicine/1949/summary.

21 Berhorst, R. "Lobotomie: Operation mit dem Eispickel," *GEO Kompakt* 2008, 15:126–131.

22 Zalashik, R. and Davidovitch, N. "Last Resort? Lobotomy Operations in Israel, 1946–60," *History of Psychiatry* 2006, 17:91–106.

23 https://www.scinexx.de/dossierartikel/klebrigenervenbahnen.

24 Biegler, R. et al. "A larger hippocampus is associated with longer-lasting spatial memory," *PNAS USA* 2001, 98:69416944.

25 Neshat-Doost, HT et al. "Reduced Specificity of Emotional Autobiographical Memories Following Self-Regulation Depletion," *Emotion* 2008, 8:731–736.

26 Tyng, CM et al. "The Influences of Emotion on Learning and Memory," *Frontiers in Psychology* 2017, www.ncbi.nlm.nih.gov/pmc/articles/PMC5573739.

27 Hilts, *Memory's Ghost*, 119.

28 Huizenga, HM et al. "Formal Models of 'Resource Depletion,'" *Behavioral and Brain Sciences* 2013, 36:694–695.

29 Martin, SB et al. "Human experience seeking correlates with hippocampus volume: Convergent evidence from manual tracing and voxel-based morphometry," *Neuropsychologia* 2007, 45:2874–2881.

30 Wheeler, JA. "Information, physics, quantum: The search for links," Proceedings III International Symposium on Foundations of Quantum Mechanics, Tokyo, 1989, 354368; https://philpapers.org/archive/WHEIPQ.pdf.

31 Toyabe, T. et al. "Experimental demonstration of information-to-energy conversion and validation of the generalized Jarzynski equality," *Nature Physics* 2010, 6:988–992.

CHAPTER 4

1 Mårtensson, J. et al. "Growth of language-related brain areas after foreign language learning," *Neuroimage* 2012, 63:240–244.

2   Abramov E et al. "Amyloidβ as a positive endogenous regulator of release probability at hippocampal synapses," *Nature Neuroscience* 2009, 12:15671576; Puzzo D. "Aβ oligomers: role at the synapse," Aging 2019, 11:10771078.

3   Yavas, E. et al. "Interactions between the hippocampus, prefrontal cortex, and amygdala support complex learning and memory," F1000Res 2019, www.ncbi.nlm.nih.gov/pmc/articles/PMC6676505.

4   Kumar, D. et al. "Sparse Activity of Hippocampal Adult-Born Neurons during REM Sleep Is Necessary for Memory Consolidation," *Neuron* 2020, 107:552–565.

5   Sadowski, JHLP et al. "Ripples Make Waves: Binding structured activity and plasticity in hippocampal networks," *Neural Plasticity* 2011, 1:1–11.

6   O'Neill, J. et al. "Play it again: Reactivation of waking experience and memory," *Trends in Neuroscience* 2010, 33:220229; Payne, JD et al. "Sleep preferentially enhances memory for emotional components of scenes," *Psychological Science* 2008, 19:781788.

7   Mendelsohn, AR and Larrick, JW. "Sleep facilitates clearance of metabolites from the brain: glymphatic function in aging and neurodegenerative diseases," *Rejuvenation Research* 2013, 16:518523.

8   Spano, GM et al. "Sleep deprivation by exposure to novel objects increases syn apse density and axonspine interface in the hippocampal CA1 region of adolescent mice," *Journal of Neuroscience* 2019, 39:66136625.

9   Lau, A. and Tymianski, M. "Glutamate receptors, neurotoxicity and neurodegeneration," *Pflugers Archiv* 2010, 460:525542.

10   MacDonald, CJ et al. "Hippocampal 'time cells' Bridge the GAP in memory for discontiguous events," *Neuron* 2011, 71: 737–749; Rolls, ET and Mills, P. "The Generation of Time in the Hippocampal Memory System," *Cell Reports* 2019, 28:1649–1658.

11   Constantinescu, AO et al. "Organizing conceptual knowledge in humans with a grid-like code," *Science* 2016, 352:1464–1468.

12   Bellmund, JLS et al. "Navigating cognition: Spatial codes for human thinking," *Science* 2018, www.ncbi.nlm.nih.gov/pubmed/30409861.

13   Gu, Y. et al. "Neurogenesis and hippocampal plasticity in adult brains," *Current Topics in Behavioral Neurosciences* 2013, 15:3148.

14   Alam, MJ et al. "Adult Neurogenesis Conserves Hippocampal Memory Capacity," *Journal of Neuroscience* 2018, 38:6854–6863.

CHAPTER 5

1   Ramón y Cajal, S. *Degeneration and Regeneration of the Nervous System*, Haffner Publishing: New York, 1928, 2:750.

2   Gross, CG. "Neurogenesis in the adult brain: Death of a Dogma," *Nature Reviews Neuroscience* 2000, 1:67–73; Colucci- D'Amato, L. et al. "The end of the central dogma of neurobiology: Stem cells and neurogenesis in adult CNS," *Neurological Sciences* 2006, 27:266–270.

3   Freund, J. et al. "Emergence of individuality in genetically identical mice," *Science* 2013, 340:756–759.

4   Altman. J. and Das, GD. "Autoradiographic and histological studies of postnatal hippocampal neurogenesis in rats," *Journal of Comparative Neurology* 1965, 124: 319–335.

5   Spalding, KL et al. "Dynamics of hippocampal neurogenesis in adult humans," *Cell* 2013, 153:1219–1227.

6   Lahdenperä, M. et al. "Fitness benefits of prolonged post-reproductive lifespan in women," *Nature* 2004, 428:178–181.

7   Lahdenperä, M. et al. "Selection for long lifespan in men: benefits of grandfathering?" *Proceedings: Biological Sciences* 2007, 274:2437–2444.

8   Gurven, M. and Kaplan, H. "Longevity among hunter- gatherers: A cross-cultural examination," *Population and Development Review* 2007, 33:321–365; Hawkes, K. "How grandmother effects plus individual variation in frailty shape fertility and mortality: Guidance from human-chimpanzee comparison," *PNAS USA* 2010, 107:8977–8984.

9   Dupret, D. et al. "Spatial relational memory requires hippocampal adult neurogenesis," *PLoS One* 2008, www.ncbi.nlm.nih.gov/pmc/articles/PMC2396793.

10 Gonçalves, JT et al. "Adult neurogenesis in the hippocampus: From stem cells to behavior," *Cell* 2016, 167:897914.

11 Toda, T. and Gage, FH. "Adult neurogenesis contributes to hippocampal plasticity," *Cell Tissue Research* 2018, 373:693709.

12 Marín-Burgin, A. and Schinder, AF. "Requirement of adult-born neurons for Hippocampus-dependent learning," *Behavioural Brain Research* 2012, 227:391399.

13 Sherry, DF and Hoshooley, JS. "Seasonal hippocampal plasticity in food-storing birds," *Philosophical Transactions of the Royal Society B* 2010, 365:933–943.

14 Jacobs, LF and Spencer WD. "Natural space use patterns and hippocampal size in kangaroo rats," *Brain Behavior and Evolution* 1994, 44:125132.

15 Clayton, NS and Krebs, JR. "Hippocampal growth and attrition in birds affected by experience," *PNAS USA* 1994, 91:74107414.

16 Obulesu, M. and Lakshmi, MJ. "Apoptosis in Alzheimer's disease: An understanding of the physiology, pathology and therapeutic avenues," *Neurochemical Research* 2014, 39:2301–2312.

17 Düzel, S. et al. "Structural Brain Correlates of Loneliness among Older Adults," *Science Reports* 2019, www.ncbi.nlm.nih.gov/pubmed/31537846.

18 Raz, N. et al. "Regional brain changes in aging healthy adults: general trends, individual differences and modifiers," *Cerebral Cortex* 2005, 15:1676–1689.

19 Barnes, J. et al. "A meta-analysis of hippocampal atrophy rates in Alzheimer's disease," *Neurobiology of Aging*, 2009, 30:1711–1723.

20 Nobis, L. et al. "Hippocampal volume across age: Nomograms derived from over 19,700 people in UK Biobank," *Neuroimage: Clinical* 2019, www.ncbi.nlm.nih.gov /pmc/articles/PMC6603440.

21 www.who.int/newsroom/factsheets/detail/depression (30.1.2020).

22   www.who.int/news/item/09122020whorevealsleadingcausesofdeathanddisabilityworl dwide20002019 (9.12.2020).

23   Humphreys, KL et al. "Evidence for a sensitive period in the effects of early life stress on hippocampal volume," *Developmental Science* 2019, www.ncbi.nlm.nih.gov/pub med/30471167.

24   Rubin, RD et al. "The role of the hippocampus in flexible cognition and social behavior," *Frontiers in Human Neuroscience* 2014, www.ncbi.nlm.nih.gov/pmc/articles/PMC4 179699.

CHAPTER 6

1   Rovelli, C. *Die Wirklichkeit, ist nicht so, wie sie scheint, Rowohlt*, Reinbek, 2016.

2   Small, SA et al. "Imaging hippocampal function across the human life span: Is memory decline normal or not?" *Annals of Neurology* 2002, 51:290–295.

3   Snyder, JS et al. "Adult hippocampal neurogenesis buffers stress responses and depressive behaviour," *Nature* 2011, 476:458–461.

4   Opel N et al. "Hippocampal Atrophy in Major Depression: a Function of Child hood Maltreatment Rather than Diagnosis?" *Neuropsychopharmacology* 2014, 39: 2723–2731.

5   Logue, MW et al. "Smaller Hippocampal Volume in Posttraumatic Stress Disorder: A Multisite ENIGMAPGC Study: Subcortical Volumetry Results From Posttraumatic Stress Disorder Consortia," *Biological Psychiatry* 2018, 83:244–253.

6   Hill, AS et al. "Increasing Adult Hippocampal Neurogenesis is Sufficient to Reduce Anxiety and Depression-Like Behaviors," *Neuropsychopharmacology* 2015, 40:2368–2378; Anacker, C. et al. "Hippocampal neurogenesis confers stress resilience by inhibiting the ventral dentate gyrus," *Nature* 2018, 559:98–102; Dranovsky, A. and Leonardo, ED. "Is there a role for young hippocampal neurons in adaptation to stress?" *Behavioural Brain Research* 2012, 227:371–375.

7   Lemaire V et al. Behavioural trait of reactivity to novelty is related to hippocampal neurogenesis," *European Journal of Neuroscience* 1999, 11:4006–4014.

8   Martin SB et al. "Human experience seeking correlates with hippocampus volume: convergent evidence from manual tracing and voxel-based morphometry," *Neuropsychologia* 2007, 45:2874–2881.

9   Lu, H. et al. "The Hippocampus underlies the association between self-esteem and physical health," *Scientific Reports* 2018, www.ncbi.nlm.nih.gov/pubmed/30459409.

10   Baumeister, RF et al. "Does high self-esteem cause better performance, inter personal success, happiness, or healthier lifestyles?" *Psychological Science in the Public Interest* 2003, 4, 1–44.

11   Kutlu, MG and Gould, TJ. "Effects of drugs of abuse on hippocampal plasticity and hippocampus-dependent learning and memory: contributions to development and maintenance of addiction," *Learn Mem* 2016, 23:515–533.

12   Schriber, RA et al. "Hippocampal volume as an amplifier of the effect of social context on adolescent depression," *Clinical Psychological Science* 2017, 5:632–649.

13 Connan, F. et al. "Hippocampal volume and cognitive function in anorexia nervosa," *Psychiatry Research* 2006, 146:117–125.

14 Lu, H. et al. "The hippocampus underlies the association between self-esteem and physical health," *Scientific Reports* 2018, www.nature.com/articles/s41598018 34793x.

15 Nehls, M. "Unified theory of Alzheimer's disease (UTAD): implications for prevention and curative therapy," *Molecular Psychiatry* 2016, www.ncbi.nlm.nih.gov/pmc/articles /PMC4947325.

16 Cooper, JM et al. "Neonatal Hypoxia, Hippocampal Atrophy, and Memory Impairment: Evidence of a Causal Sequence," *Cerebral Cortex* 2015, 25:1469–1476.

17 Duff, MC et al. "Hippocampal amnesia disrupts creative thinking," *Hippocampus* 2013, 23:1143–1149.

18 Joppich, A. "Vor welchen Krankheiten sich die Deutschen fürchten," *ÄrzteZeitung* 2019, www.aerztezeitung.de/Panorama/VorwelchenKrankheitensichdieDeutschenfuerchten 404634.html.

19 Lean, ME et al. "Primary care-led weight management for remission of type 2 diabetes (DiRECT): an open-label, cluster-randomised trial," *Lancet* 2018, 391:541–551.

20 Lean, ME et al. "Durability of a primary care-led weight-management intervention for remission of type 2 diabetes: 2-year results of the DiRECT open-label, cluster-randomised trial," *Lancet Diabetes & Endocrinology* 2019, 7:344–355.

21 www.aerztezeitung.de/Medizin/DiabetesHeilungeherunerwu enscht313647. html (4/16/2019; last viewed on 08/31/2021).

22 Lee, MM et al. "Major depression: a role for hippocampal neurogenesis?" *Current Topics in Behavioral Neurosciences* 2013, 14:153–1798; Egeland, M. et al. "Depletion of adult neurogenesis using the chemotherapy drug temozolomide in mice induces behavioural and biological changes relevant to depression," *Translational Psychiatry* 2017, www.ncbi.nlm.nih.gov/pubmed/28440814.

23 Santarelli, L. et al. "Requirement of hippocampal neurogenesis for the behavioral effects of antidepressants," *Science* 2003, 301:805–809; Boldrini, M. et al. "Antidepressants increase neural progenitor cells in the human hippocampus," *Neuropsychopharmacology* 2009, 34:2376–2389; Surget, A. et al. "Antidepressants recruit new neurons to improve stress response regulation," *Molecular Psychiatry* 2011, 16:1177–1188.

24 Cipriani, A. et al. "Comparative efficacy and acceptability of 21 antidepressant drugs for the acute treatment of adults with major depressive disorder: a systematic review and network meta-analysis," *Lancet* 2018, 391:1357–1366; Munk holm, K. et al. "Considering the methodological limitations in the evidence base of antidepressants for depression: a reanalysis of a network meta-analysis," *BMJ Open* 2019, www.ncbi .nlm.nih.gov/pmc/articles/PMC6597641.

25 Lim, GY et al. "Prevalence of Depression in the Community from 30 Countries between 1994 and 2014," *Scientific Reports* 2018, www.ncbi.nlm.nih.gov/pmc/artic les/PMC5809481.

26   Steffen, A. et al. "Trends in prevalence of depression in Germany between 2009 and 2017 based on nationwide ambulatory claims data," *JAD* 2020, 271:239–247.

27   WHO. The world health report 2001: Mental health: new understanding, new hope. Geneva: Office of Publications, 2001; https://apps.who.int/iris/bitstream/handle/10665/42390/WHR_2001.pdf.

28   Ettman, CK et al. "Prevalence of Depression Symptoms in US Adults Before and During the COVID19 Pandemic," *JAMA Network Open* 2020, www.ncbi.nlm.nih.gov/pmc/articles/PMC7489837.

29   Tian, R. et al. "A possible change process of inflammatory cytokines in the prolonged chronic stress and its ultimate implications for health," *Scientific World Journal* 2014, www.ncbi.nlm.nih.gov/pmc/articles/PMC4065693.

30   Uno, H. et al. "Neurotoxicity of glucocorticoids in the primate brain," *Hormones and Behavior* 1994, 28:336–348.

31   Saury, JM. "The role of the hippocampus in the pathogenesis of Myalgic Encephalomyelitis/Chronic Fatigue Syndrome (ME/CFS)," *Medical Hypotheses* 2016, 86:30–38.

32   Moriya, J. et al. "Resveratrol improves hippocampal atrophy in chronic fatigue mice by enhancing neurogenesis and inhibiting apoptosis of granular cells," *Biological and Pharmaceutical Bulletin* 2011, 34:354359.

33   Haß, U. et al. "Anti-Inflammatory Diets and Fatigue," *Nutrients* 2019, www.ncbi.nlm.nih.gov/pmc/articles/PMC6835556.

34   Lim, EJ et al. "Systematic review and meta-analysis of the prevalence of chronic fatigue syndrome/myalgic encephalomyelitis (CFS/ME)," *Journal of Translational Medicine* 2020, www.ncbi.nlm.nih.gov/pubmed/32093722.

35   Johnston, S. et al. "The prevalence of chronic fatigue syndrome/myalgic encephalo myelitis: a meta-analysis," *Clinical Epidemiology* 2013, 5:105–110.

36   Marshall, M. "The four most urgent questions about long COVID," *Nature* 2021, 594:168–170.

37   Komaroff, AL and Bateman, L. "Will COVID-19 Lead to Myalgic Encephalomyelitis /Chronic Fatigue Syndrome?" *Frontiers in Medicine* 2021, www.ncbi.nlm.nih.gov/pmc/articles/PMC7848220.

38   Broad, K. "Mechanisms and potential treatments for declining olfactory function and neurogenesis in the ageing brain," *Journal of Gerontology and Geriatrics* 2017, 65:93 100; WelgeLüssen A. "Ageing, neurodegeneration, and olfactory and gustatory loss," BENT 2009, 5:129132.

39   Höck, AD. "Review: Vitamin D3 deficiency results in dysfunctions of immunity with severe fatigue and depression in a variety of diseases," *In Vivo* 2014, 28:133–145.

40   Nehls, M. *Das CoronaSyndrom: Wie das Virus unsere Schwächen offenlegt—und wie wir uns nachhaltig schützen können*, Heyne: München, 2021.

41   Drake, TM et al. "Characterisation of in-hospital complications associated with COVID19 using the ISARIC WHO Clinical Characterisation Protocol UK: a prospective, multi-centre cohort study," *Lancet* 2021, 398:223–237.

42 GBD 2016 Dementia Collaborators. "Global, regional, and national burden of Alzheimer's disease and other dementias, 1990–2016: a systematic analysis for the Global Burden of Disease Study 2016," *Lancet Neurology* 2019, 18:88–106.

43 Popp, J. et al. "Cerebrospinal fluid cortisol and clinical disease progression in MCI and dementia of Alzheimer's type," *Neurobiology of Aging* 2015, 36:601–607.

44 Anderson, P. "Depressive Junge—demente Alte? Besonders Depression in jungen Jahren ist mit erhöhtem AlzheimerRisiko assoziiert," Medscape 2017, https://deutsch.medscape.com/artikelansicht/4906232.

45 Nehls, M. "Unified theory of Alzheimer's disease (UTAD): implications for prevention and curative therapy," *J Mol Psychiatry* 2016, www.ncbi.nlm.nih.gov/pmc/articles/PMC4947325.

46 Moreno Jiménez, EP et al. "Adult hippocampal neurogenesis is abundant in neurologically healthy subjects and drops sharply in patients with Alzheimer's disease," *Nature Medicine* 2019, 25:554–560; Tobin, MK et al. "Human hippocampal neurogenesis persists in aged adults and Alzheimer's disease patients," *Cell Stem Cell* 2019, www.ncbi.nlm.nih.gov/pmc/articles/PMC6608595.

47 Bennett, S. and Thomas, AJ. "Depression and dementia: cause, consequence or coincidence?" *Maturitas* 2014, 79:184–190; Hsiao, JJ and Teng, E. "Depressive symptoms in clinical and incipient Alzheimer's Disease," *Neurodegenerative Disease Management* 2013, 3:147–155.

48 Jack, CR Jr. et al. "Brain beta-amyloid measures and magnetic resonance imaging atrophy both predict time to progression from mild cognitive impairment to Alzheimer's disease," *Brain* 2010, 133:33363348.

49 Nyberg, L. et al. "Educational attainment does not influence brain aging," *PNAS USA* 2021, www.ncbi.nlm.nih.gov/pmc/articles/PMC8106299.

50 Martin, S. "Diabetes Heilung eher unerwünscht!" *ÄrzteZeitung* 2019, www.aerztezeitung.de/Medizin/DiabetesHeilungeherunerwue nscht313647.html.

51 KKHStressAuswertung: "Immer mehr Sechs bis 18Jährige psychischkrank," October 24, 2018, www.kkh.de/presse/fokusthemen/endstationdepression.

52 "Kinder und Jugendgesundheit an Schulen," School year 2017–2018, www.dak.de/dak/download/ergebnisbericht2090980.pdf.

53 Davis, MH. "Measuring individual differences in empathy: Evidence for a multidimensional approach," *Journal of Personality and Social Psychology* 1983, 44:113–126.

54 Ceballos, G. et al. "Biological annihilation via the ongoing sixth mass extinction signaled by vertebrate population losses and declines," *PNAS USA* 2017, 114:6089–6096.

55 Rice, D. "'Biological annihilation:' Earth's 6th mass extinction is under way," *USA Today*, July 10, 2017, https://eu.usatoday.com/story/tech/science/2017/07/10/earthfacessixthmassextinction/465655001/(12/16/2019; last viewed at 08/30/2021).

56 Rapoport, JL et al. "Neurodevelopmental model of schizophrenia: update 2012," *Molecular Psychiatry* 2012, 17:1228–1238.

57 Nelson, MD et al. "Hippocampal volume reduction in schizophrenia as assessed by magnetic resonance imaging: a metaanalytic study," *Archives of General Psychiatry* 1998, 55:433–440.

58 Toro, C. and Deakin, J. "Adult neurogenesis and schizophrenia: a window on abnormal early brain development?" Schizophr Res 2007, 90:1–14; Kempermann, G. et al. "The contribution of failing adult hippocampal neurogenesis to psychiatric disorders," *Current Opinion in Psychiatry* 2008, 21:290–295; Reif, A. et al. "Neurogenesis and schizophrenia: dividing neurons in a divided mind?" *European Archives of Psychiatry and Clinical Neuroscience* 2007, 257:290–299.

59 Arain, M. et al. "Maturation of the adolescent brain," *Neuropsychiatric Disease and Treatment* 2013, 9:449–461.

60 Kinney, DK et al. "Relation of schizophrenia prevalence to latitude, climate, fish consumption, infant mortality, and skin color: a role for prenatal vitamin D deficiency and infections?" *Schizophrenia Bulletin* 2009, 35:582–595.

61 Pawełczyk, T. et al. "Omega-3 fatty acid supplementation may prevent loss of gray matter thickness in the left parietooccipital cortex in first episode schizophrenia: A secondary outcome analysis of the OFFER randomized controlled study," *Schizophrenia Research* 2017, www.ncbi.nlm.nih.gov/pubmed/29079060.

62 Mossaheb, N. et al. "Polyunsaturated fatty acids in emerging psychosis," *Current Pharmaceutical Design* 2012, 18:576–591.

63 Amminger, GP et al. "Long-chain omega-3 fatty acids for indicated prevention of psychotic disorders: a randomized, placebo-controlled trial," *Archives of General Psychiatry* 2010, 67:146–154.

64 Hensch, TK. "Critical period plasticity in local cortical circuits," *Nature Reviews Neuroscience* 2005, 6:877–888; Arain, M. et al. "Maturation of the adolescent brain," *Neuropsychiatric Disease and Treatment.* 2013, 9:449–461.

65 Franke, B. et al. "The genetics of attention deficit/hyperactivity disorder in adults, a review," *Molecular Psychiatry* 2012, 17:960–987.

66 Asherson, P. et al. "Adult attention-deficit hyperactivity disorder: key conceptual issues," *Lancet Psychiatry* 2016, 3:568–578.

67 Burleson, W. "A Review of CoMorbid Depression in Pediatric ADHD: Etiologies, Phenomenology, and Treatment," *Journal of Child and Adolescent Psycho-pharmacology* 2008, 18: 565–571.

68 Xia, W. et al. "Comorbid anxiety and depression in school-aged children with attention deficit hyperactivity disorder (ADHD) and self-reported symptoms of ADHD, anxiety, and depression among parents of school-aged children with and without ADHD," *Shanghai Archives of Psychiatry* 2015, 27:356–367.

69 Liu, J. et al. "The mediating role of sleep in the fish consumption—cognitive functioning relationship: a cohort study," *Science Reports* 2017, www.ncbi.nlm.nih.gov/pubmed/29269884.

70   Hawkey, E. and Nigg, JT. "Omega-3 fatty acid and ADHD: blood level analysis and meta-analytic extension of supplementation trials," *Clinical Psychology Review* 2014, 34:496–505.

71   Gray, C. and Climie, EA. "Children with Attention Deficit/Hyperactivity Disorder and Reading Disability: A Review of the Efficacy of Medication Treatments," *Frontiers in Psychology* 2016, www.ncbi.nlm.nih.gov/pmc/articles/PMC4932103.

72   Milte, CM et al. "Eicosapentaenoic and docosahexaenoic acids, cognition, and behavior in children with attention deficit/hyperactivity disorder: A randomized controlled trial," *Nutrition* 2012, 28:670–677.

73   Yang, TX et al. "Impaired Memory for Instructions in Children with Attention Deficit Hyperactivity Disorder Is Improved by Action at Presentation and Recall," *Frontiers in Psychology* 2017, www.ncbi.nlm.nih.gov/pmc/articles/PMC5258743.

74   Woo, HD et al. "Dietary patterns in children with attention deficit/hyperactivity disorder (ADHD)," *Nutrients* 2014, 6:1539–1553.

75   Sharif, MR et al. "The Relationship between Serum Vitamin D Level and Attention Deficit Hyperactivity Disorder," *Iranian Journal of Child Neurology* 2015, 9: 48–53.

76   Derbyshire, E. "Do Omega 3/6 Fatty Acids Have a Therapeutic Role in Children and Young People with ADHD?" J Lipids 2017, www.ncbi.nlm.nih.gov/pmc/articles/PMC5603098.

77   Spikins, P. "How our autistic ancestors played an important role in human evolution," *The Conversation*, March 27, 2017, https://theconversation.com/how-our-autistic-ancestors-played-an-important-role-in-human-evolution-73477.

78   Fombonne, E.. "Epidemiology of pervasive developmental disorders," *Pediatric Research* 2009, 65:591–598.

79   van Elst K et al. "Food for thought: dietary changes in essential fatty acid ratios and the increase in autism spectrum disorders," *Neuroscience Biobehavioral Review* 2014, 45:369378.

80   Chen, J. et al. "Synaptic proteins and receptors defects in autism spectrum disorders," *Frontiers in Cellular Neuroscience* 2014, www.ncbi.nlm.nih.gov/pmc/articles/PMC4161164.

81   Meguid, NA et al. "Role of polyunsaturated fatty acids in the management of Egyptian children with autism," *Clinical Biochemistry* 2008, 41:1044–1048.

82   Canales, JJ. "Adult neurogenesis and the memories of drug addiction," *European Archives of Psychiatry and Clinical Neuroscience* 2007, 257:261–270.

83   Chambers, RA. "Adult hippocampal neurogenesis in the pathogenesis of addiction and dual diagnosis disorders," *Drug Alcohol Dependence* 2013, 130:1–12.

84   Mandyam, CD and Koob, GF. "The addicted brain craves new neurons: putative role for adult-born progenitors in promoting recovery," *Trends in Neuroscience* 2012, 35:250–260.

CHAPTER 7

1 Schumacher, YO et al. "Physiology, Power Output, and Racing Strategy of a Race Across America Finisher," *Medicine & Science in Sports and Exercise* 2011, 43: 885–889; Nehls, M. *Herausforderung Race Across America—4800 km Zeitfahren von Küste zu Küste*, Mental Enterprises: Vörstetten, 2012.

2 Nehls, M. *Das CoronaSyndrom: Wie das Virus unsere Schwächen offenlegt—und wie wir uns nachhaltig schützen können*, Heyne: München, 2021.

3 Bahnsen, U. "Was wird aus mir?" *Die Zeit* 2018, 43:33–35.

4 Flores-Tejeida, LB et al. "A review of hot and sweet pepper added in animal nutrition: alternative against the use of antibiotics," XIV International Engineering Congress (CONIIN), Queretaro 2018, https://ieeexplore.ieee.org/document/8489822.

5 Nielsen, R. et al. "Recent and ongoing selection in the human genome," *Nature Reviews Genetics* 2007, 8:857–868; Sabeti, PC et al. "Genome-wide detection and characterization of positive selection in human populations," *Nature* 2007, 449:913–918.

6 Boyce, DG et al. "Global phytoplankton decline over the past century," *Nature* 2010, 466:591–596.

7 Williams, R. "Microscopic algae produce half the oxygen we breathe," *The Science Show*, October 25, 2013, https://www.abc.net.au/listen/programs/scienceshow/microscopic-algae-produce-half-the-oxygen-we-breathe/5041338.

8 Marean, CW et al. "Als die Menschen fast ausstarben," *Spektrum der Wissenschaft* 2010, 12:59–65.

9 Marean, CW. "When the Sea Saved Humanity," *Scientific America* 2012, 22: 52–59.

10 Behar, DM et al. "The Dawn of Human Matrilineal Diversity," *American Journal of Human Genetics* 2008; Marean, CW et al. "Early human use of marine resources and pigment in South Africa during the Middle Pleistocene," *Nature* 2007, 449:905908.

11 Crawford, MA and Broadhurst, CL. "The role of docosahexaenoic and the marine food web as determinants of evolution and hominid brain development: the challenge for human sustainability," *Nutrition and Health* 2012, 21:1739.

12 Bradbury, J. "Docosahexaenoic acid (DHA): an ancient nutrient for the modern human brain," *Nutrients* 2011, 3:529–554.

13 Hill, J. et al. "Similar patterns of cortical expansion during human development and evolution," *PNAS USA* 2010, 107:13135–13140.

14 von Schacky, C. and Harris, WS. "Cardiovascular benefits of omega-3 fatty acids," *Cardiovascular Research* 2007, 73:310–315; Harris WS. "The omega-3 index as a risk factor for coronary heart disease," *American Journal of Clinical Nutrition* 2008, 87:1997–2002.

15 Kuipers, RS et al. "Estimated macronutrient and fatty acid intakes from an East African Paleolithic diet," *British Journal of Nutrition* 2010, 104:1666–1687.

16 Oken, E. et al. "Associations of maternal fish intake during pregnancy and breastfeeding duration with attainment of developmental milestones in early child hood: a study from the Danish National Birth Cohort," *American Journal of Clinical Nutrition* 2008, 88:789–796.

17   Helland, IB et al. "Maternal supplementation with very-long-chain n3 fatty acids during pregnancy and lactation augments children's IQ at 4 years of age," *Pediatrics* 2003, 111:39–44.

18   Ramakrishnan, U. et al. "Effects of docosahexaenoic acid supplementation during pregnancy on gestational age and size at birth: randomized, double-blind, placebo-controlled trial in Mexico," *Food and Nutrition Bulletin* 2010, 31:108–116.

19   Boucher, O. et al. "Neurophysiologic and neurobehavioral evidence of beneficial effects of prenatal omega-3 fatty acid intake on memory function at school age," *American Journal of Clinical Nutrition* 2011, 93:1025–1037.

20   Cohen, JT et al. "A quantitative analysis of prenatal intake of n3 polyunsaturated fatty acids and cognitive development," *American Journal of Preventive Medicine* 2005, 29:366–374.

21   Helland, IB et al. "Effect of supplementing pregnant and lactating mothers with n3 very-long-chain fatty acids on children's IQ and body mass index at 7 years of age," *Pediatrics* 2008, 122:472–479.

22   Tan ZS et al. "Red blood cell omega-3 fatty acid levels and markers of accelerated brain aging," *Neurology* 2012, 78:658664.

23   Zhang, YP et al. "Effects of DHA Supplementation on Hippocampal Volume and Cognitive Function in Older Adults with Mild Cognitive Impairment: A 12-Month Randomized, Double-Blind, Placebo-Controlled Trial," *Journal of Alzheimers Disease* 2017, 55:497–507.

24   Marean, CW. "The origins and significance of coastal resource use in Africa and Western Eurasia," *Journal of Human Evolution* 2014, 77:17–40.

25   Wilson, OE. *Die soziale Eroberung der Erde: Eine biologische Geschichte des Menschen*, C.H. Beck: München, 2016.

26   BarYosef, O. "The role of western Asia in modern human origins," *Philosophical Transactions of the Royal Society B: Biological Sciences* 1992, 337:193–200.

27   Strasser, TF et al. "Stone Age Seafaring in the Mediterranean: Evidence from the Plakias Region for Lower Palaeolithic and Mesolithic Habitation of Crete," *Hesperia* 2010, 79:145–190.

28   Brown, KS et al. "Fire as an engineering tool of early modern humans," *Science* 2009, 325:859–862.

29   Dyble, M. et al. "Engagement in agricultural work is associated with reduced leisure time among Agta hunter-gatherers," *Nature Human Behavior* 2019, 3:792–796.

30   Briker, R. and Schwenkenbecher, I. "Los! Jetzt! Schnell!" *Gehirn and Geist* 2020, 10: 12–17.

31   Ruff, CB et al. "Body mass and encephalization in Pleistocene Homo," *Nature* 1997, 387:173–176.

32   Hoffmann M. "The human frontal lobes and frontal network systems: an evolutionary, clinical, and treatment perspective," *ISRN Neurol* 2013, www.ncbi.nlm.nih.gov/pubmed/23577266.

33 Bocherens, H. et al. "Isotopic evidence for diet and subsistence pattern of the Saint Césaire I Neanderthal: review and use of a multisource mixing model," *Journal of Human Evolution* 2005, 49:71–87.

34 Bartzokis, G. et al. "Age-related changes in frontal and temporal lobe volumes in men: a magnetic resonance imaging study," *Archives of General Psychiatry* 2001, 58:461–465.

35 Richards, MP et al. "Isotope evidence for the intensive use of marine foods by Late Upper Palaeolithic humans," *Journal of Human Evolution* 2005, 49:390–394.

36 Ruff, CB et al. "Body mass and encephalization in Pleistocene Homo," *Nature* 1997, 387:173–176.

37 Pinker, S. *Das unbeschriebene Blatt: Die moderne Leugnung der menschlichen Natur*, Berlin: Fischer, 2017.

38 de Nève, D. "Religion spielt eine wichtige Rolle im Parteien-Wettbewerb," Deutschlandfunk, December 3, 2018, https://www.deutschlandfunk.de/glaube-in-der-politik-religion-spielt-eine-wichtige-rolle-100.html.

39 "Den Kohleausstieg wollen Sie doch auch, oder?" Zeit Online, January 16, 2019, https://www.zeit.de/2019/04/nachhaltigkeit-technologie-annalena-baerbock-christian-lindner.

40 Schröder, M. "Im Namen des Mammon," Tagesspiegel, December 20, 2009, https://www.tagesspiegel.de/wirtschaft/im-namen-des-mammon-6782281.html.

41 Here are only three examples, more follow in the further text: https://www.tages-spiegel.de/politik/entscheidung-von-agrarminister-schmidt-warum-das-ja-zu-glyphosat-gift-fuer-eine-moegliche-koalitionsbildung-ist/20642378.html (11/28/2017, last viewed at 08/30/2021); www.tagesspiegel.de/politik/verkehrsministerscheuerwillaussetzung derfeinstaubgren zwertediskutie ren/23904174.html (24.01.2019, zuletzt abgerufen am 30.08.2021); www.aerz tezeitung.de/praxis_wirtschaft/recht/article/980481/food watchkloecknerdie seministeringesundheitsgefaehrdend.html (01/31/2019, last viewed on 08/30/2021).

CHAPTER 8

1 Camus, A. *The Myth of Sisyphus*, Rowohlt: Reinbek, 2000.

2 Holt, J. *Gibt es alles oder nichts? Eine philosophische Detektivgeschichte*, Rowohlt: Reinbek, 2014.

3 Newman, DB et al. "The dynamics of searching for meaning and presence of meaning in daily life," *Journal of Personality and Social Psychology* 2018, 86:368–379.

4 Parihar, VK et al. "Predictable chronic mild stress improves mood, hippocampal neurogenesis and memory," *Molecular Psychiatry* 2011, 16:171–183.

5 Karp, A. et al. "Mental, physical and social components in leisure activities equally contribute to decrease dementia risk," *Dement Geriatr Cogn Disord* 2006, 21:65–73; Ertel, KA et al. "Effects of social integration on preserving memory function in a nationally representative US elderly population," *American Journal of Public Health* 2008, 98:1215–1220.

6   Luerweg, F. "Worin wir Sinn finden," Spektrum.de, May 17, 2020, https://www .spektrum.de/news/sinnsuche-worin-wir-sinn-finden/1735070.

7   Marean, CW. "When the Sea Saved Humanity," *Scientific America* 2012, 22: 52–59.

8   Muentener, P. et al. "The Efficiency of Infants' Exploratory Play Is Related to Longer-Term Cognitive Development," *Frontiers in Psychology* 2018, www.ncbi.nlm.nih .gov/pmc/articles/PMC5991261.

9   Brown, S. and Vaughan, C. *Play: How It Shapes the Brain, Opens the Imagination, and Invigorates the Soul*, Avery: New York, 2010.

10   Des Marais, S. "The Importance of Play for Adults," *PsychCentral*, November 10, 2022, https://psychcentral.com/blog/the-importance-of-play-for-adults.

11   Platte, M. "The Decline of Unstructured Play," *The Genius of Play*, https://www .thegeniusofplay.org/genius/expert-advice/articles/the-decline-of-unstructured-play .aspx#.XnfWIohKiUk, accessed August 30, 2021.

12   Hamilton, J. "Scientists Say Child's Play Helps Build A Better Brain," nprEd How Learning Happens, August 6, 2014, https://www.npr.org/sections/ed/2014 /08/06/336361277/scientists-say-childs-play-helps-build-a-better-brain.

13   Pellis, S. and Pellis, V. *The Playful Brain: Venturing to the Limits of Neuroscience*, Oneworld Publications: London, 2010.

14   Davis, KL and Montag, C. "Tribute to Jaak Panksepp (1943–2017)," *Personal Neuroscience* 2018, www.ncbi.nlm.nih.gov/pmc/articles/PMC7219686.

15   Hamilton, "Scientists Say Child's Play."

16   Pellis, SM. and Pellis, VC. "What is Play Fighting and what is it good for?" *Learning & Behavior* 2017, 45:355–366.

17   Caprara, GV et al. "Prosocial Foundations of children's academic achievement," *Psychological Science* 2000, 11:302–306.

18   Council on School Health. "The crucial role of recess in school," *Pediatrics* 2013, 131:183–188.

19   National Association for the Education of Young Children. "The value of school recess and outdoor play," Washington, DC: National Association for the Education of Young Children, 1998.

20   Committee on Physical Activity and Physical Education in the School Environment; Food and Nutrition Board; Kohl, HW III and Cook, HD (Hg.). *Educating the Student Body: Taking Physical Activity and Physical Education to School*, National Academies Press: Washington, 2013, www.ncbi.nlm.nih.gov/books/NBK201495.

21   Poulain, T. et al. "Media Use of Mothers, Media Use of Children, and Parent Child Interaction Are Related to Behavioral Difficulties and Strengths of Children," *International Journal of Environmental Research and Public Health* 2019, https: //pubmed.ncbi.nlm.nih. GOV/31766650.

22   Parisi, JM et al. "Can the wisdom of aging be activated and make a difference societally?" *Educational Gerontology* 2009, 35:867–879.

23   Carlson, MC et al. "Impact of the Baltimore Experience Corps Trial on cortical and hippocampal volumes," *Alzheimer's and Dementia Journal* 2015, 11:1340–1348.

24 Verghese, J. et al. "Leisure activities and the risk of dementia in the elderly," *New England Journal of Medicine* 2003, 348:2508–2516.

25 Stine Morrow, EAL et al. "The effects of an engaged lifestyle on cognitive vitality: A field experiment," *Psychology and Aging* 2008, 23:778–786.

26 Costa, DA et al. "Enrichment improves cognition in AD mice by amyloid-related and unrelated mechanisms," *Neurobiology and Aging* 2007, 28:831–844.

27 Lazarov, O. et al. "Environmental enrichment reduces a-beta levels and amyloid deposition in transgenic mice," *Cell* 2005, 120, 701–713; Hu, YS et al. "Complex environment experience rescues impaired neurogenesis, enhances synaptic plasticity, and attenuates neuropathology in family Alzheimer's disease-linked APP-swe/PS1 -Delta-E9 mice," *FASEB* 2010, 24: 1667–1681.

28 Praag, H van et al. "Neural consequences of environmental enrichment," *Nature Neuroscience* 2000, 1:191–198; Mirochnic, S. et al. "Age effects on the regulation of adult hippocampal neurogenesis by physical activity and environmental enrichment in the APP23 mouse model of Alzheimer disease," *Hippocampus* 2009, 19:1008–1018.

29 Lindstrom, HA et al. "The relationships between television viewing in midlife and the development of Alzheimer's disease in a case-control study," *Brain and Cognition* 2005, 58:157–165.

30 Durchschnittliche Fernsehdauer in Deutschland in den Jahren von 1997 bis 2020 in Minuten pro Tag. Statista. Das Statistikportal, 2013; http://de.statista.com/statistik /daten/studie/118/umfrage/fernsehkonsumentwic klungdersehdauerseit1997.

31 Hoang, TD et al. "Effect of Early Adult Patterns of Physical Activity and Television Viewing on Midlife Cognitive Function," *JAMA Psychiatry* 2016, 73:73–79.

32 Yerkes, RM and Dodson, JD. "The relation of strength of stimulus to rapidity of habit-formation," *Journal of Comparative Neurology and Psychology* 1908, 18:459–482.

33 Saaltink, DJ and Vreugdenhil, E. "Stress, glucocorticoid receptors, and adult neu-rogenesis: a balance between excitation and inhibition?" *Cellular and Molecular Life Sciences* 2014, 71:2499–2515.

34 Nehls M. *Alzheimer ist heilbar: Rechtzeitig zurück in ein gesundes Leben*, Heyne: München, 2015, 85.

35 Logue, MW et al. "Smaller Hippocampal Volume in Posttraumatic Stress Disorder: A Multisite ENIGMAPGC Study: Subcortical Volumetry Results From Post-traumatic Stress Disorder Consortia," *Biol Psychiatry* 2018, 83:244–253.

36 Buwalda, B. et al. "Long-term effects of social stress on brain and behavior: a focus on hippocampal functioning," *Neuroscience & Biobehavioral Reviews* 2005, 29:83–97.

37 Lindgren, L. et al. "Longitudinal Evidence for Smaller Hippocampus Volume as a Vulnerability Factor for Perceived Stress," *Cerebral Cortex* 2016, 26:3527–3533.

38 Radley, JJ et al. "Chronic behavioral stress induces apical dendritic reorganization in pyramidal neurons of the medial prefrontal cortex," *Neuroscience* 2004,125: 1–6; Liston, C., Miller, MM et al. "Stress-induced alterations in prefrontal cortical den-dritic morphology predict selective impairments in perceptual attentional set-shift-ing," *Journal of Neuroscience* 2006, 26:7870–7874.

39 Stillman, TF et al. "Alone and Without Purpose: Life Loses Meaning Following Social Exclusion," *Journal of Experimental Social Psychology* 2009, 45:686–694.

40 Williams, KD et al. "Cyber-ostracism: Effects of Being Ignored Over the Internet," *Journal of Personality and Social Psychology* 2000, 79:748–762; Sander, M. "Emotionale Reaktionen auf experimentell induzierten sozialen Ausschluss," *Diplomarbeit* 2010, www. grin.com/document/322127.

41 Frankl, VE. . . . *trotzdem Ja zum Leben sagen: Ein Psychologe erlebt das Konzentrationslager*, Penguin Verlag: München, 2018.

42 Luerweg. "Worin wir Sinn finden."

43 Wallert, M. *Stark durch Krisen: Von der Kunst, nicht den Kopf zu verlieren*, Econ: Berlin, 2020.

44 Luerweg. "Worin wir Sinn finden."

45 Luerweg. "Worin wir Sinn finden."

46 Brassai, L. et al. "A reason to stay healthy: The role of meaning in life in relation to physical activity and healthy eating among adolescents," *Journal of Health Psychology* 2015, 20:473–482.

47 Boyle, PA et al. "Effect of purpose in life on the relation between Alzheimer disease pathologic changes on cognitive function in advanced age," *Archives of General Psychiatry* 2012, 69:499–505.

48 Fredrickson, BL et al. "A functional genomic perspective on human well-being," *PNAS USA* 2013, 110:13684–13689.

49 Harris, TB et al. "Associations of elevated interleukin-6 and C-reactive protein levels with mortality in the elderly," *American Journal of Medicine* 1999, 106:506–512; Reuben, DB et al. "Peripheral blood markers of inflammation predict mortality and functional decline in high-functioning community-dwelling older persons," *Journal of the American Geriatrics Society* 2002, 50:638–644; De Martinis et al. "Inflammation markers predicting frailty and mortality in the elderly," *Experimental and Molecular Pathology* 2006, 80:219–227.

50 Goshen, I. et al. "Brain interleukin1 mediates chronic stress-induced depression in mice via adrenocortical activation and hippocampal neurogenesis suppression," Mol Psychiatry 2008, 13:717–728; Bowen, KK et al. "Adult interleukin-6 knockout mice show compromised neurogenesis," *Neuroreport* 2011, 22:126–130; Iosif, RE et al. "Tumor necrosis factor receptor 1 is a negative regulator of progenitor proliferation in adult hippocampal neurogenesis," *Journal of Neuroscience* 2006, 26:9703–9712.

51 Monje, ML et al. "Inflammatory blockade restores adult hippocampal neurogenesis," *Science* 2003, 302:1760–1765.

52 Ryff, CD et al. "Positive health: connecting well-being with biology," Philosophical Transactions of the Royal Society B: *Biological Sciences* 2004, 359:1383–1394.

53 Boccardi, M. and Boccardi, V. "Psychological Well-being and Healthy Aging: Focus on Telomeres," *Geriatrics* 2019, www.ncbi.nlm.nih.gov/pmc/articles/PMC6473912.

54 Schaefer, SM et al. "Purpose in life predicts better emotional recovery from negative stimuli," *PLoS One* 2013, www.ncbi.nlm.nih.gov/pmc/articles/PMC3827458; Zilioli,

S. et al. "Purpose in life predicts allostatic load ten years later," *Journal of Psychosomatic Research* 2015, 79:451–457.

55   Luerweg, F.: "Worin wir Sinn finden", Spektrum.de, www.spektrum.de/news/sinnsuche -worin-wir-sinn-finden/1735070 (17.05.2020, last accessed 30.08.2021).

CHAPTER 9

1   Lifei, Liu L. et al. "Progesterone increases rat neural progenitor cell cycle gene expression and proliferation via extracellularly regulated kinase and progesterone receptor membrane components 1 and 2," *Endocrinology* 2009, 150:3186–3196; Zhang, Z. et al. "Progesterone promotes the survival of newborn neurons in the dentate gyrus of adult male mice," *Hippocampus* 2010, 20:402–412; Sheppard PAS et al. "Structural plasticity of the hippocampus in response to estro gens in female rodents," *Molecular Brain* 2019, www.ncbi.nlm.nih.gov/pmc/articles/PMC6423800; Lin, YT et al. "Oxytocin stimulates hippocampal neurogenesis via oxytocin receptor expressed in CA3 pyramidal neurons," *Nature Communications* 2017, https://pubmed.ncbi .nlm.nih.gov/28912554; Walker, TL et al. "Prolactin stimulates precursor cells in the adult mouse hippocampus," *PLoS One* 2012, www.ncbi.nlm.nih.gov/pmc/articles /PMC3433411.

2   Leuner, B. and Sabihi, S. "The birth of new neurons in the maternal brain: Hormonal regulation and functional implications," *Frontiers in Neuroendocrinology* 2016, 41:99–113.

3   Torner, L. et al. "Prolactin prevents chronic stress-induced decrease of adult hippocampal neurogenesis and promotes neuronal fate," *Journal of Neuroscience* 2009, 29:1826–1833; Leuner, B. et al. "Oxytocin stimulates adult neurogenesis even under conditions of stress and elevated glucocorticoids," *Hippocampus* 2012, 22:861–868.

4   UvnäsMoberg, K. "Oxytocin may mediate the benefits of positive social interaction and emotions," *Psychoneuroendocrinology* 1998, 23:819–835.

5   Feldman, R. and Eidelman, AI. "Direct and indirect effects of breast milk on the neurobehavioral and cognitive development of premature infants," *Developmental Psychobiology* 2003, 43:109–119.

6   Victora, CG et al. "Breastfeeding in the 21st century: epidemiology, mechanisms, and lifelong effect," *Lancet* 2016, 387:475–490.

7   World Health Organization. "Breastfeeding," https://www.who.int/health-topics/breast feeding#tab=tab_1.

8   Victora, CG et al. "Association between breastfeeding and intelligence, educational attainment, and income at 30 years of age: a prospective birth cohort study from Brazil," *Lancet Glob Health* 2015, 3:199–205.

9   Black, RE et al. "Maternal and child undernutrition and overweight in low income and middle-income countries," Lancet 2013, 382:427–451; "Breastfeeding: achieving the new normal," *Lancet* 2016, pubmed.ncbi.nlm.nih.gov/26869549.

10 Ystrom, E. "Breastfeeding cessation and symptoms of anxiety and depression: a longitudinal cohort study," *BMC Pregnancy Childbirth* 2012, www.ncbi.nlm.nih.gov /pmc/articles/PMC3449190.

11 Petersen, I. et al. "Depression, depressive symptoms and treatments in women who have recently given birth: UK cohort study," *BMJ Open* 2018, www.ncbi.nlm.nih .gov/pmc/articles/PMC6224756.

12 Jacobs, A. "Opposition to Breast-Feeding Resolution by U.S. Stuns World Health Officials," *The New York Times*, July 8, 2018, https://www.nytimes.com/2018/07/08 /health/world-health-breastfeeding-ecuador-trump.html.

13 RuizPeláez, J. et al. "Kangaroo Mother Care, an example to follow from developing countries," *BMJ* 2004, 329:1179–1181.

14 Charpak, N. et al. "Twenty-year Follow-up of Kangaroo Mother Care Versus Traditional Care," *Pediatrics* 2017, https://pubmed.ncbi.nlm.nih.gov/27965377.

15 Ropars, S. et al. "The long-term effects of the Kangaroo Mother Care intervention on cognitive functioning: Results from a longitudinal study," *Developmental Neuropsychology* 2018, 43:82–91.

16 Chen, EM et al. "Effects of Father-Neonate Skin-to-Skin Contact on Attachment: A Randomized Controlled Trial," *Nursing Research and Practice* 2017, www.ncbi .nlm. nih.gov/pmc/articles/PMC5282438; Srinath, BK et al. "Kangaroo care by fathers and mothers: comparison of physiological and stress responses in preterm infants," *Journal of Perinatology* 2016, 36:401–404.

17 Ainsworth, MD and Bell, SM: "Attachment, exploration, and separation: illustrated by the behavior of one-year-olds in a strange situation," *Child Development* 1970, 41:49–67.

18 Smarius, LJ et al. "Excessive infant crying doubles the risk of mood and behavioral problems at age 5: evidence for mediation by maternal characteristics," *European Child & Adolescent Psychiatry* 2017, 26:293–302.

19 Moore, ER et al. "Early skin-to-skin contact for mothers and their healthy new born infants," *Cochrane Database Syst Rev.* 2012, https://www.cochranelibrary.com/cdsr /doi/10.1002/14651858.CD003519.pub4/full.

20 Schölmerich, A. "Culture and early infancy among central African foragers and farmers," *Developmental Psychology* 1998, 34:653–661; Konner, M. *Hunter-Gatherer Childhoods*, Routledge: Oxford, UK 2017, Chapter "Hunter-gatherer infancy and childhood: The! Kung and others," 19–64.

21 Sarah, R. Moore SR et al. "Epigenetic correlates of neonatal contact in humans," *Developmental Psychopathology* 2017, 29:1517–1538.

22 Arai, JA et al. "Transgenerational rescue of a genetic defect in long-term potentiation and memory formation by juvenile enrichment," *Journal of Neuroscience* 2009, 29: 1496–1502.

23 Unternaehrer, E. et al. "Childhood maternal care is associated with DNA methylation of the genes for brain-derived neurotrophic factor (BDNF) and oxytocin receptor (OXTR) in peripheral blood cells in adult men and women," *Stress* 2015, 18:451–461.

24   Unternaehrer, E. et al. "Dynamic changes in DNA methylation of stress-associated genes (OXTR, BDNF) after acute psychosocial stress," *Translational Psychiatry* 2012, https://pubmed.ncbi.nlm.nih.gov/22892716.

25   McGowan, PO et al. "Epigenetic regulation of the glucocorticoid receptor in human brain associates with childhood abuse," *Nature Neuroscience* 2009, 12:342–348.

26   Sultan, FA and Day, JJ. "Epigenetic mechanisms in memory and synaptic function," *Epigenomics* 2011, 3:157–181.

27   Arai, JA and Feig, LA. "Long-lasting and transgenerational effects of an environmental enrichment on memory formation," *Brain Research Bulletin* 2001, 85:30–35; Gräff, J. et al. "An epigenetic blockade of cognitive functions in the neurodegenerating brain," *Nature* 2012, 483:222–226.

28   Koch, J. "Das Leben vor der Geburt," *Der Spiegel*, June 17, 2012, https://www.spiegel.de/politik/das-leben-vor-der-geburt-a-ac890e08–0002-0001–0000-000086505890.

29   Luby, JL et al. "Maternal support in early childhood predicts larger hippocampal volumes at school age," *PNAS USA* 2012, 109:2854–2859; Luby, JL et al. "Preschool is a sensitive period for the influence of maternal support on the trajectory of hippocampal development," *PNAS USA* 2016, 113:5742–5747.

30   Tomoda, A. et al. "Reduced prefrontal cortical gray matter volume in young adults exposed to harsh corporal punishment," *Neuroimage* 2009, 47:66–71.

31   Woon, FL and Hedges, DW. "Hippocampal and amygdala volumes in children and adults with childhood maltreatment-related posttraumatic stress disorder: a meta-analysis," *Hippocampus* 2008, 18:729–736; Riem, MM et al. "Beating the brain about abuse: Empirical and meta-analytic studies of the association between maltreatment and hippocampal volume across childhood and adolescence," *Developmental Psychopathology* 2015, 27:507–520.

32   www.tagesschau.de/inland/kriminalstatistikkinder101.html (5/26/2021).

33   www.deutschlandfunk.de/kriminalstatistik-gewalt-gegen-kinder-erneut-gestiegen .1769.de.html?dram:article_id=391032;www.tagesschau.de/inland/kriminalstatistik -kindesmissbrauchsfaelle-103.html; www.deutschlandfunkkultur.de/unicef-studie-gewalt -gegen-kinder-kommt-in-al len-schichten.1008.de.html?dram:article_id=296604.

34   Leuner, B. et al. "Sexual experience promotes adult neurogenesis in the hippocampus despite an initial elevation in stress hormones," *PloS One* 2010, www. ncbi.nlm.nih .gov/pubmed/20644737.

35   Wang, H. et al. "Histone deacetylase inhibitors facilitate partner preference formation in female prairie voles," *Nature Neuroscience* 2013, 16:919–924.

36   Scheele, D. et al. "Oxytocin modulates social distance between males and females," *Journal of Neuroscience* 2012, 32:16074–16079.

37   Kaplan, HS and Robson, AJ. "The emergence of humans: The coevolution of intelligence and longevity with intergenerational transfers," *PNAS USA* 2002, 99:10221–10226.

38   Kosfeld, M. et al. "Oxytocin increases trust in humans," *Nature* 2005, 435:673–676.

39   Boyd, R. et al. "The cultural niche: why social learning is essential for human adaptation," *PNAS USA* 2011, 108:10918–10925; Pinker, S. "Colloquium paper: the cognitive niche: coevolution of intelligence, sociality, and language," *PNAS USA* 2010, 107:8993–8999.

40   De Dreu, C. "Oxytocin modulates cooperation within and competition between groups: an integrative review and research agenda," *Hormones and Behavior* 2012, 61:419–428.

41   Asch, SE. "Effects of group pressure upon the modification and distortion of judgment," in Guetzkow, HS, *Groups, Leadership and Men: Research in Human Relations*, Carnegie Press: Pittsburgh, 1951; Asch, SE. "Studies of independence and conformity: I. A minority of one against a unanimous majority," *Psychological Monographs* 1956, 70:1–70.

42   Voland, E. *Die Natur des Menschen. Grundkurs Soziobiologie*, C.H. Beck: München, 2007, 37.

43   Liebal, K. and Call, J. "The origins of nonhuman primates' manual gestures," *Philosophical Transactions of the Royal Society London B: Biological Science* 2012, 367:118–128; Steele, J. et al. "From action to language: comparative perspectives on primate tool use, gesture and the evolution of human language," *Philosophical Transactions of the Royal Society London B: Biological Science* 2012, 367:4–9.

44   Simpson, EA et al. "The mirror neuron system as revealed through neonatal imitation: presence from birth, predictive power and evidence of plasticity," *Philosophical Transactions of the Royal Society London B: Biological Science* 2014, https://pubmed.ncbi.nlm.nih.gov/24778381.

45   Giganti F and Esposito ZM. "Contagious and spontaneous yawning in autistic and typically developing children," *Current Psychology Letters* 2009, www.ncbi.nlm.nih.gov/pmc/articles/PMC3736493.

46   Usui, S. et al. "Presence of contagious yawning in children with autism spectrum disorder," *Autism Research and Treatment* 2013, www.ncbi.nlm.nih.gov/pmc/articles/PMC3736493.

47   Glocker, ML et al. "Baby schema modulates the brain reward system in nulliparous women," *PNAS USA* 2009, 106:9115–9119.

48   Nagasawa, M. et al. "Social evolution. Oxytocin-gaze positive loop and the coevolution of human-dog bonds," *Science* 2015, 348:333–336; Borgi, M. and Cirulli, F. "Pet Face: Mechanisms Underlying Human-Animal Relationships," *Frontiers in Psychology* 2016, www.ncbi.nlm.nih.gov/pmc/articles/PMC4782005.

49   Maute, M. et al. "Der Blick zum Säugling—gestört durch Smartphones?" *Obstetrica* 2018, https://www.zhaw.ch/storage/psychologie/upload/forschung/psychotherapie/smart-toddlers/Maute_et_al_vonWyl_2018.pdf.

50   Andrews, S. et al. "Beyond Self Report: Tools to Compare Estimated and Real-World Smartphone Use," *PLoS One* 2015, https://pubmed.ncbi.nlm.nih.gov/26509895.

51   Herrmann, S. "Mach doch mal das Handy aus," Suddeutsche Zeitung, June 21, 2018, https://www.sueddeutsche.de/wissen/eltern-und-ihre-smartphones-mach-mal-aus-mama-1.4025658?reduced=true.

52   www.bkz.de/nachrichten/handykonsumvonelternschadet babysaufklae runggefordert 72112.html (05/4/2020, last viewed on 09/9/2021).

53   Fombonne, E. "Epidemiology of pervasive developmental disorders," Pediatric Research 2009, 65:591–598.

54   Cacioppo, JT et al. "Social isolation," Annals of the New York Academy of Sciences 2011, 123:17–22.

55   Kirsch, P. "Oxytocin in the socioemotional brain: implications for psychiatric disorders," Dialogues in Clinical Neuroscience 2015, 17:463–476.

56   Wismer Fries, AB et al. "Early experience in humans is associated with changes in neuropeptides critical for regulating social behavior," PNAS USA 2005, 102:17237–17240.

57   Mumtaz, F. et al. "Neurobiology and consequences of social isolation stress in animal model—A comprehensive review," Biomedicine & Pharmacotherapy 2018, 105:1205–1222.

58   SaumAldehoff, T. "Im Gefängnis der Einsamkeit," Psychologie Heute 2012, 7:61–66.

59   Cinini, SM et al. "Social isolation disrupts hippocampal neurogenesis in young nonhuman primates," Frontiers in Neuroscience 2014, https://pubmed.ncbi.nlm.nih. gov/24733997.

60   Luerweg, F. "Epigenetik: Psychotherapie für die Gene?" Gehirn and Geist 2020, https://www.spektrum.de/news/epigenetik-psychotherapie-fuer-die-gene/1737168.

61   Akbaraly, TN et al. "Leisure activities and the risk of dementia in the elderly," Neurology 2009, 73: 854–861; Takeda, T. et al. "Psychosocial risk factors involved in progressive dementia-associated senility among the elderly residing at home. AGES project-three year cohort longitudinal study," Nihon Koshu Eisei Zasshi 2010, 57:1054–1065; Saczynski, JS et al. "The effect of social engagement on incident dementia: the Honolulu-Asia Aging Study," American Journal of Epidemiology 2006, 163:433–440; Fratiglioni, L. et al. "Influence of social network on occurrence of dementia: a community-based longitudinal study," Lancet 2000, 355:1315–1319; Zunzunegui, M. et al. "Social networks, social integration, and social engagement determine cognitive decline in community-dwelling Spanish older adults," Journal of Gerontology 2003, 58:93–100.

62   Ertel, KA et al. "Effects of social integration on preserving memory function in a nationally representative US elderly population," American Journal of Public Health 2006, 98:1215–1220.

63   Scaccianoce S et al. "Social isolation selectively reduces hippocampal brain-derived neurotrophic factor without altering plasma corticosterone," Behavioural Brain Research 2006, 168:323325.

64   Alberti, FB. "This 'Modern Epidemic': Loneliness as an Emotion Cluster and a Neglected Subject in the History of Emotions," Emotion Review, https://doi.org/10.1177/1754073918768876.

65 The Federal Government's repsonse to the small question from some members of parliament and the GFDP parmiliamentary group—Drucksache 19/9880—05/23/2019, "Einsamkeit und die Auswirkung auf die öffentliche Gesundheit," https://dip21 .bundestag.de/dip21/btd/19/104/1910456.pdf.

66 Holt-Lunstad, J. et al. "Social relationships and mortality risk: a meta-analytic review," *PLoS Med* 2010, https://pubmed.ncbi.nlm.nih.gov/20668659.

67 www.ageuk.org.uk/globalassets/age-uk/documents/reports-and-publications/reports -and-briefings/active-communities/rb_dec17_jocox_commission_finalreport.pdf.

68 www.nytimes.com/2018/01/17/world/europe/ukbritainloneli ness.html.

69 Banerjee, D. and Rai, M. "Social isolation in Covid19: The impact of loneliness," *International Journal of Social Psychiatry* 2020, 66:525–527; Killgore, WDS et al. "Loneliness: A signature mental health concern in the era of COVID19," Psychiatry Res 2020, www.ncbi.nlm.nih.gov/pmc/articles/PMC7255345.

70 van Tilburg, TG et al. "Loneliness and mental health during the COVID19 pandemic: A study among Dutch older adults," *Journals of Gerontology* 2020, https://pubmed .ncbi.nlm.nih.gov/32756931; Krendl AC and Perry BL. "The impact of sheltering-in -place during the COVID-19 pandemic on older adults' social and mental well-being," *Journals of Gerontology* 2020, https://pubmed.ncbi.nlm.nih.gov/32778899; Ornell, F. et al. "'Pandemic fear' and COVID19: mental health burden and strategies," *Braz J Psychiatry* 2020, 42:232–235.

71 Sher, L. "The impact of the COVID19 pandemic on suicide rates," *Q JM* 2020, https://doi.org/10.1093/qjmed/hcaa202; Tanaka, T. and Okamoto, S. "Increase in suicide following an initial decline during the COVID19 pandemic in Japan," *Nature Human Behavior* 2021, 5:229–238; Sánchez GR. "Monthly suicide rates during the COVID19 pandemic: Evidence from Japan," *Economics Letters* 2021, www .sciencedirect.com/science/article/pii/S0165176521002913; Thakur, V and Jain, A. "COVID 2019–Suicides: A global psychological pandemic," *Brain, Behavior, and Immunity* 2020, 88:952–953.

72 Hommel, T. "Kinderärzte fordern Öffnung von Kitas und Schulen," *ArzteZeitung*, February 10, 2021, https://www.aerztezeitung.de/Politik/Corona-Gespraeche -Kinderaerzte-fordern-Oeffnung-von-Kitas-und-Schulen-416972.html.

73 Heu, LC et al. "Lonely Alone or Lonely Together? A Cultural-Psychological Examination of Individualism-Collectivism and Loneliness in Five European Countries," *Personality and Social Psychology Bulletin* 2019, 45:780–793; Beller, J and Wagner, A. "Loneliness and Health: The Moderating Effect of Cross-Cultural Individualism/Collectivism," *Journal of Aging and Health* 2020, https://pubmed .ncbi.nlm.nih.gov/32723203.

74 Lemola, S. et al. "Adolescents' electronic media use at night, sleep disturbance, and depressive symptoms in the smartphone age," *Journal of Youth and Adolescence* 2015, 44:405–418.

75   Wilmer, HH et al. "Smartphones and Cognition: A Review of Research Exploring the Links between Mobile Technology Habits and Cognitive Functioning," *Frontiers in Psychology* 2017, www.ncbi.nlm.nih.gov/pmc/articles/PMC5403814.

76   Hutton, JS et al. "Associations Between Screen-Based Media Use and Brain White Matter Integrity in Preschool-Aged Children," *JAMA Pediatr* 2020, www.ncbi.nlm.nih.gov/pmc/articles/PMC6830442.

CHAPTER 10

1   Williamson, AM and Feyer, AM. "Moderate sleep deprivation produces impairments in cognitive and motor performance equivalent to legally prescribed levels of alcohol intoxication," *Occup Environ Med* 2000, 57:649–655.

2   Kuo, AA. "Resident overnight call—an idea past its time?" *West J Med* 2001, 174:180–181.

3   Lovinger, DM et al. "Ethanol inhibits NMDA-activated ion current in hippocampal neurons," *Science* 1989, 243:1721–1724.

4   Ramírez Rodríguez, G. et al. "Melatonin modulates cell survival of new neurons in the hippocampus of adult mice," *Neuropsychopharmacology* 2009, 34:2180 2191; Vander Weele, CM et al. "Restoration of hippocampal growth hormone reverses stress-induced hippocampal impairment," *Frontiers in Behavioral Neuroscience* 2013, https://pubmed.ncbi.nlm.nih.gov/23785317.

5   Sportiche, N. et al. "Sustained sleep fragmentation results in delayed changes in hippocampaldependent cognitive function associated with reduced dentate gyrus neurogenesis," *Neuroscience* 2010, 170:247–258.

6   Kohn, M. et al. "Sleep recalibrates homeostatic and associative synaptic plasticity in the human cortex," *Nature Communications* 2016, https://pubmed.ncbi.nlm.nih.gov/27551934; Fattinger, S. et al. "Deep sleep maintains learning efficiency of the human brain," *Nature Communications* 2017, www.ncbi.nlm.nih.gov/pmc/articles/PMC5458149.

7   Fernandes, C. et al. "Detrimental role of prolonged sleep deprivation on adult neurogenesis," *Frontiers in Cellular Neuroscience* 2015, https://pubmed.ncbi.nlm.nih.gov/25926773.

8   Kurth, S. et al. "Increased Sleep Depth in Developing Neural Networks: New Insights from Sleep Restriction in Children," *Frontiers Human Neuroscience* 2016, www.ncbi.nlm.nih.gov/pmc/articles/PMC5030292.

9   Kerber, B. "Die übermüdete Gesellschaft," *Psychologie Heute* 2001, 8:60–67.

10   Nolting, HD: "Deutschland schläft schlecht – ein unterschätztes Problem," *Gesundheitsreport* 2017, www.iges.com/e6/e1621/e10211/e15829/e20067/e20074/e20075/attr_objs20081/IGES_Hans_Dieter_Nolting_15032017_ger.pdf.

11   Middleton, B. et al. "Human circadian rhythms in constant dim light (8 lux) with knowledge of clock time," *Journal of Sleep Research* 1996, 5:69–76; Bonmati Carrion, MA et al. "Living Without Temporal Cues: A Case Study," *Frontiers in Physiology* 2020, https://pubmed.ncbi.nlm.nih.gov/32116739.

12 Chattu, VK et al. "The Global Problem of Insufficient Sleep and Its Serious Public Health Implications," *Healthcare* (Basel) 2018, www.ncbi.nlm.nih.gov/pmc/articles /PMC6473877.

13 Isaiah, A. et al. "Associations between frontal lobe structure, parent-reported obstructive sleep disordered breathing and childhood behavior in the ABCD dataset," *Nature Communications* 2021, www.nature.com/articles/s41467021225340.

14 Novati, A. et al. "Chronic sleep restriction causes a decrease in hippocampal volume in adolescent rats, which is not explained by changes in glucocorticoid levels or neurogenesis," *Neuroscience* 2011, 190:145–155.

15 Zit.n. Shafy, S. "Der Nachtkampf," *Der Spiegel* 2011, 44:130–139, www.spiegel.de /spiegel/print/d81303024.html.

16 Sundelin, T. et al. "Negative effects of restricted sleep on facial appearance and social appeal," *Royal Society Open Science* 2017, www.ncbi.nlm.nih.gov/pmc/articles /PMC5451790.

17 Ben Simon, E. and Walker, MP. "Sleep loss causes social withdrawal and loneliness," *Nature Communications* 2018, www.ncbi.nlm.nih.gov/pmc/articles/PMC6092357.

18 Gordon AM et al. "The social side of sleep: Elucidating the links between sleep and social processes," *Current Directions in Psychological Science* 2017, 26:470–475.

19 Taveras, EM et al. "Prospective Study of Insufficient Sleep and Neurobehavioral Functioning Among School-Age Children," *Academic Pediatrics* 2017, 17:625–632.

20 Kidwell, KM et al. "Stimulant Medications and Sleep for Youth With ADHD: A Meta-analysis," *Pediatrics* 2015, 136:1144–1153.

21 Smith, P. "Hitlers geheime Drogensucht," *ÄrzteZeitung*, October 24, 2016, https: //www.aerztezeitung.de/Medizin/Hitlers-geheime-Drogensucht-311969.html.

22 Matricciani, L. et al. "In search of lost sleep: secular trends in the sleep time of school-aged children and adolescents," *Sleep Medicine Reviews* 2012, 16:203–211.

23 www.dak.de/dak/bundesthemen/fastjederdritteschuelerh atschlafstoerun gen2090982 .html.

24 Trauner, S. "Wenn Schüler zu wenig schlafen," *ÄrzteZeitung*, January 15, 2019, https://www.aerztezeitung.de/Medizin/Wenn-Schueler-zu-wenig-schlafen-254178 .html.

25 www.chip.de/news/BesserschlafengesuenderarbeitenDieseDownloads-schonen-Ihre -Augen_88486348.html.

26 Saikhedkar, N. et al. "Effects of mobile phone radiation (900 MHz radiofrequency) on structure and functions of rat brain," *Neurology Research* 2014, 36:1072–1079; Miller, AB et al. "Risks to Health and Well-Being From Radio-Frequency Radiation Emitted by Cell Phones and Other Wireless Devices," *Front Public Health* 2019, www .ncbi.nlm.nih.gov/pmc/articles/PMC6701402.

27 Lee, H. et al. "Effects of exercise with or without light exposure on sleep quality and hormone responses," *Journal of Exercise Nutrition & Biochemistry* 2014, 18:293–299.

28 Carskadon, MA et al. "Association between puberty and delayed phase preference," *Sleep* 1993, 16:258–262.

29   Institute of Medicine (US) Committee on Sleep Medicine and Research; Colten, HR and Altevogt, BM (Hg.). *Sleep Disorders and Sleep Deprivation: An Unmet Public Health Problem*, National Academies Press: Washington, DC, 2006, third chapter: "Extent and Health Consequences of Chronic Sleep Loss and Sleep Disorders," https://www.ncbi.nlm.nih.gov/books/NBK19961/.

30   Farrell, P. et al. "Shedding light on daylight saving time," J.P. Morgan Chase Institute 2016, https://www.jpmorganchase.com/content/dam/jpmc/jpmorgan-chase-and-co/institute/pdf/jpmc-institute-daylight-savings-report.pdf.

31   Zhang, H. et al. "Measurable health effects associated with the daylight saving time shift," *PLoS Comput Biol* 2020, www.ncbi.nlm.nih.gov/pmc/articles/PMC7302868.

32   Poteser, M. and Moshammer, H. "Daylight Saving Time Transitions: Impact on Total Mortality," *International Journal of Environmental Research and Public Health* 2020, www.ncbi.nlm.nih.gov/pmc/articles/PMC7084938.

33   Zerbini, G. et al. "Lower school performance in late chronotypes: underlying factors and mechanisms," *Scientific Reports* 2017, https://pubmed.ncbi.nlm.nih.gov/28663569.

34   Alfonsi, V. et al. "Later School Start Time: The Impact of Sleep on Academic Performance and Health in the Adolescent Population," *International Journal of Environmental Research and Public Health* 2020, https://pubmed.ncbi.nlm.nih.gov/32283688.

35   Barnes, M. et al. "Setting Adolescents Up for Success: Promoting a Policy to Delay High School Start Times," *Journal of School Health* 2016, 86:552–557.

36   Antúnez, JM. "Circadian typology is related to emotion regulation, metacognitive beliefs and assertiveness in healthy adults," *PLoS One* 2020, https://pubmed.ncbi.nlm.nih.gov/32168366.

37   Kuperczkó, D. et al. "Late bedtime is associated with decreased hippocampal volume in young healthy subjects," *Sleep and Biological Rhythms* 2015, 13:68–75.

38   Norbury, R. "Chronotype, depression and hippocampal volume: cross-sectional associations from the UK Biobank," Chronobiol Int 2019, 36:709–716; Partonen, T. "Chronotype and Health Outcomes," *Current Sleep Medicine Reports* 2015, 1:205–211/.

39   Knutson, KL and von Schantz, M. "Associations between chronotype, morbidity and mortality in the UK Biobank cohort," *Chronobiol Int* 2018, 35:1045–1053.

40   Feriante, J. and Singh, S. "REM Rebound Effect," StatPearls, www.ncbi.nlm.nih.gov/books/NBK560713.

41   Waters, F. et al. "Severe Sleep Deprivation Causes Hallucinations and a Gradual Progression Toward Psychosis With Increasing Time Awake," *Frontiers in Psychiatry* 2018, https://pubmed.ncbi.nlm.nih.gov/30042701.

42   Roehrs, T. and Roth, T. "Drug-related Sleep Stage Changes: Functional Significance and Clinical Relevance," *Sleep Medicine Clinics* 2010, 5:559–570.

43   www.dak.de/dak/download/gesundheitsreport2017–2108948.pdf.

44 McMillan, JM et al. "Management of insomnia and long-term use of sedative hypnotic drugs in older patients," *CMAJ* 2013, 185:1499–1505.

45 Guina J and Merrill B. "Benzodiazepines I: Upping the Care on Downers: The Evidence of Risks, Benefits and Alternatives," *Journal of Clinical Medicine* 2018, www.ncbi.nlm.nih.gov/pmc/articles/PMC5852433.

46 He Q et al. "Risk of Dementia in Long-Term Benzodiazepine Users: Evidence from a Meta-Analysis of Observational Studies," *Journal of Clinical Neurology* 2019, 15:919.

47 Whitney, P. et al. "Feedback Blunting: Total Sleep Deprivation Impairs Decision Making that Requires Updating Based on Feedback," *Sleep* 2015, 38:745–754.

48 Alkozei, A. et al. "Chronic Sleep Restriction Increases Negative Implicit Attitudes Toward Arab Muslims," *Scientific Reports* 2017, www.ncbi.nlm.nih.gov/pmc/articles/PMC5487318.

49 Muraven, M. "Prejudice as self-control failure," *Journal of Applied Social Psychology* 2008, 38:314–333.

50 Ghumman, S. and Barnes, CM. "Sleep and prejudice: a resource recovery approach," *Journal of Applied Social Psychology* 2013, 166–178.

51 Pepin, E. et al. "Shift work, night work and sleep disorders among pastry cookers and shopkeepers in France: a cross-sectional survey," *BMJ Open* 2018; Strohmaier, S. et al. "A Review of Data of Findings on Night Shift Work and the Development of DM and CVD Events: a Synthesis of the Proposed Molecular Mechanisms," *Current Diabetes Reports* 2018, https://pubmed.ncbi.nlm.nih.gov/30343445.

52 von Leszczynski, U. "Der frühe Vogel—Schlaflose Nächte, übermüdete Nation," *ÄrzteZeitung*, June 19, 2018, https://www.aerztezeitung.de/Medizin/Der-Albtraum-vom-Nicht-Schlafen-225817.html.

53 Dunietz, GL et al. "Later School Start Times: What Informs Parent Support or Opposition?" *Journal of Clinical Sleep Medicine* 2017, 13:889–897.

CHAPTER 11

1 https://adipositas-gesellschaft.de/deutsche-adipositas-gesellschaft-zum-weltadipositas-tag-am-04-03-2020.

2 Pulgaron, ER and Delamater, AM. "Obesity and type 2 diabetes in children: epidemiology and treatment," *Current Diabetes Reports* 2014, www.ncbi.nlm.nih.gov/pmc/artic les/PMC4099943.

3 www.who.int/newsroom/factsheets/detail/obesity-and-overweight (last accessed on 13.9.2021; this page is continually updated).

4 Hales, CM et al. "Prevalence of Obesity Among Adults and Youth: United States, 2015–2016," NCHS Data Brief 2017, No. 288:1–8, https://pubmed.ncbi.nlm.nih.gov/29155689.

5 Schienkiewitz, A. et al. "Body mass index among children and adolescents: prevalences and distribution considering underweight and extreme obesity: Results of KiGGS Wave 2 and trends," *Bundesgesundheitsblatt Gesundheitsforschung Gesundheitsschutz* 2019, 62:1225–1234; Nittari, G. et al. "Fighting obesity in children from European

World Health Organization member states. Epidemiological data, medical-social aspects, and prevention programs," *Clinical Therapeutics* 2019, 170:223–230.

6   Rolland Cachera, MF et al. "Early adiposity rebound: causes and consequences for obesity in children and adults," *International Journal of Obesity* 2006, 4:11–17; Evensen, E. et al. "Tracking of overweight and obesity from early childhood to adolescence in a population-based cohort—the Tromsø Study, Fit Futures," *BMC Pediatrics* 2016, www.ncbi.nlm.nih.gov/pmc/articles/PMC4863357.

7   OrtizPinto, MA et al. "Association between general and central adiposity and development of hypertension in early childhood," *European Journal of Preventive Cardiology* 2019, 26:1326–1334.

8   Sarganas, G. et al. "Tracking of Blood Pressure in Children and Adolescents in Germany in the Context of Risk Factors for Hypertension," *International Journal of Hypertension* 2018, www.ncbi.nlm.nih.gov/pmc/articles/PMC6178151.

9   Schaare, HL et al. "Association of peripheral blood pressure with gray matter volume in 19- to 40-year-old adults," *Neurology* 2019, 92:758–773.

10 NCD Risk Factor Collaboration (NCDRisC). "Trends in adult body-mass index in 200 countries from 1975 to 2014: a pooled analysis of 1698 population-based measurement studies with 19.2 million participants," *Lancet* 2016, 387:1377–1396.

11 www.aerzteblatt.de/nachrichten/73523/UngesundesEssen-verschuldet-jaehr-lich-mehr-als-400–000-kardiovaskulaere-Todesfaelle (3/13/2017, last viewed on 08/30/2021).

12 Röckl, S. et al. "All-cause mortality in adults with and without type 2 diabetes: findings from the national health monitoring in Germany," *BMJ Open Diabetes Research & Care* 2017, www.ncbi.nlm.nih.gov/pmc/articles/PMC5759714.

13 Martin, S. "Mehr Respekt für die Ernährungsberatung!," www.aerztezeitung.de /Medizin/Mehr-Respekt-fuer-die-Ernaehrungsberatung-224403.html (18.5.2018, last accessed on 31.08.2021); Schaff, T. "So hoch ist die Diabetes-Prävalenz in der Welt," www.aerztezeitung.de/Medizin/So-hoch-ist-die-Diabetes-Praevalenz-in-der-Welt -403972.html (11/15/2019, last viewed at 08/31/2021).

14 Dawkins, R. *The Selfish Gene: 40th Anniversary edition*, Oxford University Press: Oxford, 2016.

15 Nehls, M. *Die MethusalemStrategie. Vermeiden, was uns daran hindert, gesund älter und weiser zu werden*, Mental Enterprises: Vörstetten, 2011.

16 Hendrie, GA et al. "Defining the complexity of childhood obesity and related behaviours within the family environment using structural equation modelling," *Public Health Nutrition* 2011, 15:48–57.

17 Elder, JP et al. "Individual, family, and community environmental correlates of obesity in Latino elementary school children," *Journal of School Health* 2010, 80:20–30.

18 Effertz, T. and Wilcke, AC. "Do television food commercials target children in Germany?" *Public Health Nutrition* 2012, 15:1466–1473.

19 Norman, J. et al. "Sustained impact of energy-dense TV and online food advertising on children's dietary intake: a specialized, randomized, crossover, counterbalanced trial," *International Journal of Behavioral Nutrition and Physical Activity* 2018,

https://pubmed.ncbi.nlm.nih.gov/29650023; Eppinger, U. "Fernsehen animiert zu süßen Drinks, sitzend zu lernen jedoch nicht so sehr," https://deutsch.medscape.com /arti kel/4900525 (11/15/2012, last viewed at 08/31/2021).

20 Emond, JA et al. "Randomized Exposure to Food Advertisements and Eating in the Absence of Hunger Among Preschoolers," *Pediatrics* 2016, https://pediatrics . aappublications.org/content/138/6/e20162361.

21 Urbanek, M. "Wie Influencer die Ernährung von Kindern beeinflussen," www .aerztezeitung.de/Wirtschaft/Wie-Influencer-die-Ernaehrung-von-Kindern-beeinflussen -417161.html (2/17/2021, last viewed on 08/31/2021).

22 Hall, KD et al. "UltraProcessed Diets Cause Excess Calorie Intake and Weight Gain: An Inpatient Randomized Controlled Trial of Ad Libitum Food Intake," *Cell Metabolism* 2019, 30:67–77.

23 Twig, G. et al. "Body-Mass Index in 2.3 Million Adolescents and Cardiovascular Death in Adulthood," *New England Journal of Medicine* 2016, 374:2430–2440.

24 Srour, B. et al. "Ultra-processed food intake and risk of cardiovascular disease: prospective cohort study (NutriNetSanté)," *BMJ* 2019, www.ncbi.nlm.nih.gov/pmc /articles/PMC6538975; Etminan, M. "Association between consumption of ultra-processed foods and all-cause mortality: SUN prospective cohort study," *BMJ* 2019, www.bmj.com/content/365/bmj.l1949.

25 Raether, E. and Stelzer, T. "Süße Geschäfte," *Die Zeit*, www.zeit.de/2013/20/kinder -marketing-werbung (05/8/2013, last viewed on 08/31/2021).

26 Maniam, J. et al. "Sugar Consumption Produces Effects Similar to Early Life Stress Exposure on Hippocampal Markers of Neurogenesis and Stress Response," *Frontiers in Molecular Neuroscience* 2016, www.ncbi.nlm.nih.gov/pmc/articles /PMC4717325.

27 www.foodwatch.org/fileadmin/user_upload/Dok_2_201 80502_offener_Brief_AErzteschaft _Praevention_Fehlernaehrung.pdf.

28 Joppich, A. "KinderlebensmittelWerbung verbieten!," www.aerztezeitung.de/Politik /Werbung-fuer-Kinderlebensmittel-verbieten-403058.html (10/13/2019, last viewed on 08/31/2021).

29 www.aerztezeitung.de/Medizin/Mediziner-fordern-von-Politikern-staerkeren-Kampf -gegen-Uebergewicht-297605.html (5/19/2017, last viewed on 08/31/2021).

30 Ebd.

31 World Health Organization. "Taxes on sugary drinks: Why do it?" 2017, https://iris .who.int/bitstream/handle/10665/260253/WHO-NMH-PND-16.5Rev.1-eng.pdf.

32 Christian Schmidt: "Totalverbote sind verfassungsrechtlich bedenklich", www.tagesspiegel .de/wirtschaft/bundesernaehrungsminister-zu-werbeverbotentotalverbote-sind -verfassungsrechtlich-bedenklich/12312050.html (9/14/2015, last viewed on 08/31 /2021).

33 Swinburn, BA et al. "The Global Syndemic of Obesity, Undernutrition, and Climate Change: The Lancet Commission report," *Lancet* 2019, 393:791–846.

34   Stillerman, KP. "'Big Food' Companies Spend Big Money in Hopes of Shaping the Dietary Guidelines for Americans," *Union of Concerned Scientists*, June 6, 2019, https://blog.ucsusa.org/karen-perry-stillerman/big-food-companies-spend-big-money-in-hopes-of-shaping-the-dietary-guidelines-for-americans/.

35   Wallenfels, M. "Diese Ministerin ist gesundheitsgefährdend," *ÄrzteZeitung online*, www.aerztezeitung.de/Wirtschaft/Diese-Ministerin-ist-gesundheitsge-faehrdend-253189.html (01/31/2019, last viewed on 08/31/2021).

36   www.bmel.de/SharedDocs/Downloads/DE/_Ministerium/Beiraete/agrarpolitik/wbae-gutachten-nachhaltige-ernaehrung.pdf?_blob=publicationFile&v=3, (accessed on 06/9/2020).

37   www.foodwatch.org/de/pressemitteilungen/2020/foodwatch-zu-den-empfehlungen-des-wissenschaftlichen-beirates-julia-kloeckner-versagt-bei-demschutz-von-kindern.

38   www.derpostillon.com/2015/11/geschaftsmaigesterbehilfeverboten.html (11/6/2015, last viewed on 08/31/2021).

39   Meier, T. et al. "Cardiovascular mortality attributable to dietary risk factors in 51 countries in the WHO European Region from 1990 to 2016: a systematic analysis of the Global Burden of Disease Study," *European Journal of Epidemiology* 2019, 34:37–55.

40   Soreca, I. et al. "Gain in adiposity across 15 years is associated with reduced gray matter volume in healthy women," *Psychosomatic Medicine* 2009, 71:485–490.

41   Ho, AJ et al. "The effects of physical activity, education, and body mass index on the aging brain," *Human Brain Mapping* 2011, 32:1371–1382.

42   Debette, S. et al. "Visceral fat is associated with lower brain volume in healthy middle-aged adults," *Annals of Neurology* 2010, 68:136–144.

43   Ambikairajah, A. et al. "Longitudinal Changes in Fat Mass and the Hippocampus," *Obesity* 2020, 28:1263–1269.

44   Myers, MG Jr. et al. "Obesity and leptin resistance: distinguishing cause from effect," *Trends in Endocrinology & Metabolism* 2010, 21:643–651.

45   Tezapsidis, N. et al. "Leptin: a novel therapeutic strategy for Alzheimer's disease," *Journal of Alzheimer's Disease* 2009, 16:731–740; Garza, JC et al. "Leptin restores adult hippocampal neurogenesis in a chronic unpredictable stress model of depression and reverses glucocorticoid-induced inhibition of GSK3β/βcatenin signaling," *Molecular Psychiatry* 2012, 17:790–808; Greco, SJ et al. "Leptin reduces pathology and improves memory in a transgenic mouse model of Alzheimer's disease," *Journal of Alzheimer's Disease* 2010, 19:1155–1167.

46   Cherbuin, N. et al. "Being overweight is associated with hippocampal atrophy: the PATH Through Life Study," *International Journal of Obesity* 2015, 39:1509–1514.

47   Spiegel, K et al. "Brief communication: Sleep curtailment in healthy young men is associated with decreased leptin levels, elevated ghrelin levels, and increased hunger and appetite," *Annals of Internal Medicine* 2004, 141:846–850; Taheri, S. et al. "Short sleep duration is associated with reduced leptin, elevated ghrelin, and increased body mass index," *PLoS Med* 2004, https://pubmed.ncbi.nlm.nih.gov/15602591.

48   Xenaki, N. et al. "Impact of a stress management program on weight loss, mental health and lifestyle in adults with obesity: a randomized controlled trial," *Journal of Molecular Biochemistry* 2018, 7:78–84.

49   Whitmer, RA et al. "Central obesity and increased risk of dementia more than three decades later," *Neurology* 2008, 71:1057–1964.

50   Hassing, LB et al. "Overweight in midlife and risk of dementia: A 40-year follow-up study," *International Journal of Obesity* (Lond) 2009, 33:893–898; Xu, WL et al. "Midlife overweight and obesity increase late-life dementia risk. A population-based twin study," *Neurology* 2011, 76:1568–1574.

51   Sawatsky, L. et al. "Type 2 diabetes in a four-year-old child," *CMAJ* 2017, 189: 888–890.

52   Cherbuin N et al. "Higher normal fasting plasma glucose is associated with Hippocampal atrophy: The PATH Study," *Neurology* 2012, 79:1019–1026; Kerti L et al. "Higher glucose levels associated with lower memory and reduced hippocampal microstructure," *Neurology* 2013, 81:1746–1752.

53   Chaudhuri, J. et al. "The Role of Advanced Glycation End Products in Aging and Metabolic Diseases: Bridging Association and Causality," *Cell Metabolism* 2018, 28:337–352.

54   Derk, J. et al. "The Receptor for Advanced Glycation End products (RAGE) and Mediation of Inflammatory Neurodegeneration," *Journal of Alzheimer's Disease & Parkinsonism* 2018, www.ncbi.nlm.nih.gov/pmc/articles/PMC6293973; Goshen, I. et al. "Brain interleukin1 mediates chronic stress-induced depression in mice via adrenocortical activation and hippocampal neurogenesis suppression," *Molecular Psychiatry* 2008, 13:717–728; Bowen, KK et al. "Adult interleukin-6 knockout mice show compromised neurogenesis," *NeuroReport* 2011, 22:126130; Iosif, RE et al. "Tumor necrosis factor receptor 1 is a negative regulator of progenitor proliferation in adult hippocampal neurogenesis," *Journal of Neuroscience* 2006, 26:9703–9712.

55   Attuquayefio, T. et al. "A four-day Western-style dietary intervention causes reductions in hippocampal-dependent learning and memory and interoceptive sensitivity," *PLoS One* 2017, www.ncbi.nlm.nih.gov/pmc/articles/PMC5322971.

56   Guilbaud, A. et al. "How Can Diet Affect the Accumulation of Advanced Glycation End-Products in the Human Body?" *Foods* 2016, www.ncbi.nlm.nih.gov/pmc /articles/PMC5302422.

57   Hoffman, R. and Gerber, M. "Food Processing and the Mediterranean Diet," *Nutrients* 2015, 7:7925–7964.

58   Stevenson, RJ et al. "Hippocampal-dependent appetitive control is impaired by experimental exposure to a Western-style diet," *Royal Society Open Science* 2020, www .ncbi. nlm.nih.gov/pmc/articles/PMC7062097.

59   https://ec.europa.eu/info/sites/default/files/food-farming-fisheries/plants_and _plant_products/documents/report-plant-proteins-com2018–757-final_en.pdf; www .spglobal.com/platts/en/market-insights/latest-news/agriculture/060820-eu-data -soybean-meal-imports-in-2019–20-rise-2-on-year-soybean-importsup-1 (8.6.2020).

60   www.destatis.de/DE/Themen/Branchen-Unternehmen/Landwirtschaft
      -Forstwirtschaft-Fischerei/Produktionsmethoden/aktuell-duengen.html.

61   Greenpeace Nachrichten: "Gefährliche Brühe aus Tierfabriken," 2019, 01:8–12,
      https://gpn.greenpeace.de/ausgabe/01–19/gefaehrliche-bruehe-aus-tierfabriken.

62   Meier, T. et al. "Cardiovascular mortality attributable to dietary risk factors in 51
      countries in the WHO European Region from 1990 to 2016: a systematic analy-
      sis of the Global Burden of Disease Study," *European Journal of Epidemiology* 2019,
      34:37–55.

63   Ward, MH et al. "Drinking Water Nitrate and Human Health: An Updated Review,"
      International Journal of Environmental Research and Public Health 2018, www.ncbi.
      nlm.nih.gov/pubmed/30041450; de la Monte, SM et al. "The 20-Year Voyage Aboard
      the Journal of Alzheimer's Disease: Docking at 'Type 3 Diabetes,' Environmental/
      Exposure Factors, Pathogenic Mechanisms, and Potential Treatments," *Journal of
      Alzheimer's Disease* 2018, 62:1381–1390.

64   Poore, J. and Nemecek, T. "Reducing food's environmental impacts through producers
      and consumers," *Science* 2018, 360:987–992, https://www.science.org/doi/10.1126
      /science.aaq0216.

65   Cotter, DG et al. "Obligate role for ketone body oxidation in neonatal metabolic
      homeostasis," *Journal of Biological Chemistry* 2011, 286:6902–6910.

66   Maffezzini, C. et al. "Metabolic regulation of neurodifferentiation in the adult brain,"
      Cellular Molecular Life Sciences 2020, 77:2483–2496.

67   Newman, JC and Verdin, E. "Ketone bodies as signaling metabolites," *Trends in
      Endocrinology & Metabolism 2014*, 25:42–52, www.ncbi.nlm.nih.gov/pubmed
      /24140022; Mattson MP et al. "Intermittent metabolic switching, neuroplasticity and
      brain health," *Nature Reviews Neuroscience* 2018, 19:63–80.

68   Gano, LB et al. "Ketogenic diets, mitochondria, and neurological diseases," *Journal of
      Lipid Research* 2014, 55:2211–2228.

69   de la Monte, SM. "Type 3 diabetes is sporadic Alzheimer's disease: minireview,"
      European Neuropsychopharmacology 2014, 24:1954–1960.

70   Marcus, C. et al. "Brain PET in the diagnosis of Alzheimer's disease," *Clinical Nuclear
      Medicine* 2014, 39:413–426.

71   Hertz, L. et al. "Effects of ketone bodies in Alzheimer's disease in relation to neural hypo-
      metabolism, β-amyloid toxicity, and astrocyte function," *Journal of Neurochemistry*
      2015, 134:7–20.

72   Ptomey, LT et al. "Breakfast Intake and Composition Is Associated with Superior
      Academic Achievement in Elementary Schoolchildren," *Journal of the American
      College of Nutrition* 2016, 35:326–333.

73   Edefonti, V. et al. "The effect of breakfast composition and energy contribution on
      cognitive and academic performance: a systematic review," *The American Journal of
      Clinical Nutrition* 2014, 100:626–656.

74   Heyman MB et al. "Fruit Juice in Infants, Children, and Adolescents: Current
      Recommendations," *Pediatrics* 2017, https://pubmed.ncbi.nlm.nih.gov/28562300.

75 Wilkinson, MJ et al. "Ten-Hour Time-Restricted Eating Reduces Weight, Blood Pressure, and Atherogenic Lipids in Patients with Metabolic Syndrome," *Cell Metabolism* 2020, 31:92–104.

76 Harries, HC. "The evolution, dissemination and classification of Cocos nucifera," *Botanical Review* 1978, 44:265–320.

77 Hilditch, TP and Meara, ML. "Human milk fat: 1. Component fatty acids," *Biochemical Journal* 1944, 38:2934; Mazzocchi, A. et al. "The Role of Lipids in Human Milk and Infant Formulae," *Nutrients* 2018, www.ncbi.nlm.nih.gov/pmc /articles/PMC5986447.

78 Hachey, DL et al. "Human lactation. II: Endogenous fatty acid synthesis by the mammary gland," *Pediatric Research* 1989, 25:63–68.

79 Cunnane, SC and Crawford, MA. "Energetic and nutritional constraints on infant brain development: implications for brain expansion during human evolution," *Journal of Human Evolution* 2014, 77:88–98.

80 Taylor, MK et al. "Dietary Neuro-ketotherapeutics for Alzheimer's Disease: An Evidence Update and the Potential Role for Diet Quality," *Nutrients* 2019, www .ncbi.nlm.nih.gov/pmc/articles/PMC6722814.

81 Nehls, M. *Das CoronaSyndrom: Wie das Virus unsere Schwächen offenlegt—und wie wir uns nachhaltig schützen können*, Heyne: München, 2021, 132–136.

82 Lai, SM et al. "Toxic effect of acrylamide on the development of hippocampal neurons of weaning rats," *Neural Regeneration Research* 2017, 12:1648–1654; Moumtaz, S. et al. "Toxic aldehyde generation in and food uptake from culinary oils during frying practices: peroxidative resistance of a monounsaturate-rich algae oil," *Scientific Reports* 2019, www.ncbi.nlm.nih.gov/pmc/articles/PMC6412032.

83 Schneider, C. et al. "Autoxidative transformation of chiral omega6 hydroxy linoleic and arachidonic acids to chiral 4-hydroxy-2-nonenal," *Chemical Research in Toxicology* 2004, 17:937–941; Zárate, J. et al. "A study of the toxic effect of oxidized sunflower oil containing 4-hydroperoxy-2-nonenal and 4-hydroxy-2-nonenal on cortical Trk-A receptor expression in rats," *Nutritional Neuroscience* 2009, 12:249–259.

84 Nehls, M. "Kokosöl sei reines Gift—eine wissenschaftliche Gegendarstel lung," www .michaelnehls.de/kokosoel.htm (8/29/2018, last viewed on 08/31/2021).

85 Hamley, S. "The effect of replacing saturated fat with mostly n6 polyunsaturated fat on coronary heart disease: a meta-analysis of randomised controlled trials," *Nutrition Journal* 2017, www.ncbi.nlm.nih.gov/pmc/articles/PMC5437600; Vijaya kumar, M. et al. "A randomized study of coconut oil versus sunflower oil on cardiovascular risk factors in patients with stable coronary heart disease," *Indian Heart Journal 2016*, 68:498–506.

86 Dehghan, M. et al. "Prospective Urban Rural Epidemiology (PURE) study investigators. Associations of fats and carbohydrate intake with cardiovascular disease and mortality in 18 countries from five continents (PURE): a prospective cohort study," *Lancet* 2017, 390:2050–2062.

87 Astrup, A. et al. "Saturated Fats and Health: A Reassessment and Proposal for Food-Based Recommendations: JACC State of the Art Review," *Journal of the American College of Cardiology* 2020, 76:844–857; dazu ein Kommentar in der ÄrzteZeitung: www.aerztezeitung.de/Medizin/Sind-gesaettigte-Fettsaeuren-doch-gesund-414124 .html.

88 www.who.int/news/item/14–05-2018-who-plan-to-eliminate-industrially-produced -trans-fatty-acids-from-global-food-supply.

89 DGE:wissenschaft/weitere-publikationen/achinformationen/trans-fettsaeuren (11/27 /2020, last viewed at 08/31/2021).

90 Ebd.

91 Black, RE al. "Maternal and child undernutrition and overweight in low-income and middle-income countries," *Lancet* 2013, 382: 427–451.

92 Stevens, GA et al. "Trends and mortality effects of vitamin A deficiency in children in 138 low-income and middle-income countries between 1991 and 2013: a pooled analysis of population-based surveys," *Lancet Global Health* 2015, 3:528–536.

93 I.L.A. Kollektiv (Hg.). *Auf Kosten Anderer?: Wie die imperiale Lebensweise ein gutes Leben für alle verhindert*, OekomVerlag: München, 2017.

94 UNICEF, WHO and World Bank Group. "Joint malnutrition estimates," 2020 edition, https://data.unicef.org/resources/jme-report-2020/.

95 UNICEF. "Situation tracking for COVID19 socioeconomic impacts," https://data .unicef.org/resources/rapidsituationtrackingcovid19socioeconomicimpactsdataviz /(last viewed in July 2021).

96 Laborde, D. et al. "Poverty and food insecurity could grow dramatically as COVID19 spreads," IFPRI, April 16, 2020, https://www.ifpri.org/blog/poverty-and-food-insecurity -could-grow-dramatically-covid-19-spreads/.

97 www.welthungerhilfe.de/fileadmin/pictures/publications/de/position_papers/2021co vid19diecoronapandemieverschaerftdenh ungerweltweit.pdf.

98 World Food Programme. "COVID19 will double number of people facing food crises unless swift action is taken," 2020, https://www.africa-newsroom.com/press /coronavirus-africa-covid19-will-double-number-of-people-facing-food-crises-unless -swift-action-is-taken?lang=en.

99 Gamba, C. et al. "Genome flux and stasis in a five millennium transect of European prehistory," *Nature Communications* 2014, www.ncbi.nlm.nih.gov/pmc/articles/PMC 4218962.

100 Pyörälä, S. "Treatment of mastitis during lactation," *Irish Veterinary Journal* 2009, 62:40–44.

101 Saed, HAER and Ibrahim, HMM. "Antimicrobial profile of multi-drug-resistant Streptococcus spp. isolated from dairy cows with clinical mastitis," *Journal of Advanced Veterinary and Animal Research* 2020, 7:186–197.

102 Prestinaci F et al. "Antimicrobial resistance: a global multifaceted phenomenon," *Pathogens and Global Health* 2015, 109:309–318; Martin MJ et al. "Antibiotics

Overuse in Animal Agriculture: A Call to Action for Health Care Providers," *American Journal of Health* 2015, 105:2409–2410.

103 www.who.int/news/item/29–04-2019-new-report-calls-for-urgent-action-toavert -antimicrobial-resistance-crisis (29.4.2019).

104 Martin. "Antibiotics Overuse in Animal Agriculture."

105 Flom, JD and Sicherer, SH. "Epidemiology of Cow's Milk Allergy," *Nutrients* 2019, www.ncbi.nlm.nih.gov/pmc/articles/PMC6566637.

106 zur Hausen, H. et al. "Specific nutritional infections early in life as risk factors for human colon and breast cancers several decades later," *International Journal of Cancer* 2019, 144:1574–1583.

107 www.dkfz.de/de/presse/download/HintergrundPKPlasmido me_final.pdf.

108 Triantis, V. et al. "Immunological Effects of Human Milk Oligosaccharides," *Frontiers in Pediatrics* 2018, www.ncbi.nlm.nih.gov/pmc/articles/PMC6036705.

109 El Hodhod MA et al. "Cow's Milk Allergy Is a Major Contributor in Recurrent Perianal Dermatitis of Infants," *ISRN Pediatrics* 2012, www.ncbi.nlm.nih.gov/pmc /articles/PMC3439954.

110 Kahn, A. et al. "Insomnia and cow's milk allergy in infants," *Pediatrics* 1985, 76: 880–884.

111 Michaëlsson K et al. "Milk intake and risk of mortality and fractures in women and men: cohort studies," *BMJ* 2014, https://pubmed.ncbi.nlm.nih.gov/25352269.

112 Hilliard, CB. "High osteoporosis risk among East Africans linked to lactase persistence genotype," *BoneKEy* Reports 2016, https://pubmed.ncbi.nlm.nih. gov/27408710.

113 Zhang, Q. et al. "D-galactose injured neurogenesis in the hippocampus of adult mice," *Neurological Research* 2005, 27:552–556; Cui, X. et al. "Chronic systemic D-galactose exposure induces memory loss, neurodegeneration, and oxidative damage in mice: protective effects of Ralphalipoic acid," *Journal of Neuroscience Research* 2006, 83:1584–1590.

114 Gao, J. et al. "Salidroside suppresses inflammation in a D-galactose-induced rat model of Alzheimer's disease via SIRT1/NFκB pathway," *Metabolic Brain Disease* 2016, 31:771–778; Li, L. et al. "Moderate exercise prevents neurodegeneration in D-galactose-induced aging mice," *Neural Regeneration Research* 2016, 11:807–815.

115 Gu, Y. et al. "Food combination and Alzheimer disease risk: a protective diet," *Arch Neurol* 2010, 67:699–706.

116 Johnson, DL et al. "Breast feeding and children's intelligence," *Psychol Rep* 1996, 79:1179–1185; Isaacs, EB et al. "Impact of breast milk on intelligence quotient, brain size, and white matter development," *Pediatric Research* 2010, 67:357–362.

117 Sommerfeld, M. et al. "Trans unsaturated fatty acids in natural products and processed foods," *Progress in Lipid Research* 1983, 22:221–233; Pfalzgraf, A. et al. "Gehalte an transFettsäuren in Lebensmitteln," *Z Ernährungswiss* 1993, 33:24–43.

118 Gebauer, SK et al. "Vaccenic acid and trans fatty acid isomers from partially hydrogenated oil both adversely affect LDL cholesterol: a double-blind, randomized controlled trial," *American Journal of Clinical Nutrition* 2015, 102:1339–1346;

Brouwer, IA et al. "Effect of animal and industrial trans fatty acids on HDL and LDL cholesterol levels in humans—a quantitative review," *PLoS One* 2010, www.ncbi .nlm.nih.gov/pmc/articles/PMC2830458.

119　Chen, M. et al. "Dairy fat and risk of cardiovascular disease in 3 cohorts of US adults," *American Journal of Clinical Nutrition* 2016, 104:1209–1217.

120　Melnik, BC and Schmitz, G. "Milk's Role as an Epigenetic Regulator in Health and Disease," *Diseases* 2017, www.ncbi.nlm.nih.gov/pmc/articles/PMC5456335.

121　Lamb, M. et al. "The effect of childhood cow's milk intake and HLADR genotype on risk of islet autoimmunity and type 1 diabetes: the Diabetes Autoimmunity Study in the young," *Pediatric Diabetes* 2015, 16: 31–33; Chia, JSJ et al. "A1 beta-casein milk protein and other environmental predisposing factors for type 1 diabetes," *Nutrition & Diabetes* 2017, www.ncbi.nlm.nih.gov/pmc/articles/PMC5518798.

122　Niinistö, S. et al. "Fatty acid status in infancy is associated with the risk of type 1 diabetes-associated autoimmunity," *Diabetologia* 2017, https://pubmed.ncbi.nlm .nih.gov/28474159.

123　zur Hausen, H. "Risk factors: What do breast and CRC cancers and MS have in common?" *Nature Reviews Clinical Oncology* 2015, 12:569–570.

124　Norris, JM et al. "Omega-3 polyunsaturated fatty acid intake and islet autoimmunity in children at increased risk for type 1 diabetes," *JAMA* 2007, 298:1420–1428.

125　Mennella, JA et al. "Effects of cow milk versus extensive protein hydrolysate formulas on infant cognitive development," *Amino Acids* 2016, 48:697–705.

126　Souter I et al. "The association of protein intake (amount and type) with ovarian antral follicle counts among infertile women: results from the EARTH prospective study cohort," *BJOG* 2017, 124:1547–1555.

127　Schaum, J. et al. "A national survey of persistent, bio-accumulative, and toxic (PBT) pollutants in the United States milk supply," *Journal of Exposure Analysis and Environmental Epidemiology* 2003, https://pubmed.ncbi.nlm.nih.gov/12743612; Liao, C. and Kannan, K. "Concentrations and profiles of bisphenol A and other bisphenol analogues in foodstuffs from the United States and their implications for human exposure," *Journal of Agricultural and Food Chemistry* 2013, 61:4655–4662.

128　Maruyama, K. et al. "Exposure to exogenous estrogen through intake of commercial milk produced from pregnant cows," *Pediatrics International* 2010, 52:3338; Ganmaa D et al. "Milk, dairy intake and risk of endometrial cancer: a 26-year follow-up," *International Journal of Cancer* 2012, 130:2664267; Melnik BC et al., "The impact of cow's milk-mediated mTORC-1 signaling in the initiation and progression of prostate cancer," *Nutrition & Metabolism* 2012, https://pubmed.ncbi .nlm.nih.gov/22891897; Torfadottir JE et al. "Milk intake in early life and risk of advanced prostate cancer," *American Journal of Epidemiology* 2012, www.ncbi.nlm .nih.gov/pmc/articles/PMC3249408.

129　Stork, R. "Alles wie gehabt," Spektrum, www.spektrum.de/kolumne/alleswiegehabt /1793111 (11/13/2020, last viewed on 08/31/2021).

130 Jacka, FN et al. "Western diet is associated with a smaller hippocampus: a longitudinal investigation," *BMC Med* 2015, https://pubmed.ncbi.nlm.nih. gov/26349802.

131 Orlich, MJ et al. "Vegetarian dietary patterns and mortality in Adventist Health Study 2," *JAMA Intern Med* 2013, 173:1230–1238.

132 Sarter, B. et al. "Blood docosahexaenoic acid and eicosapentaenoic acid in vegans: Associations with age and gender and effects of an algal-derived omega-3 fatty acid supplement," *Clin Nutr* 2015, 34:212–218; Kornsteiner, M. et al. "Very low n3 long-chain polyunsaturated fatty acid status in Austrian vegetarians and vegans," *Annals of Nutrition and Metabolism* 2008, 52:37–47.

133 Jia, S. et al. "Elevation of Brain Magnesium Potentiates Neural Stem Cell Proliferation in the Hippocampus of Young and Aged Mice," *Journal of Cellular Physiology* 2016, 231:1903–1912.

134 Markhus, MW et al. "Low omega-3 index in pregnancy is a possible biological risk factor for postpartum depression," *PLoS One* 2013, www.ncbi.nlm.nih.gov/pmc /articles/PMC3701051.

135 Golding, J. et al. "High levels of depressive symptoms in pregnancy with low omega-3 fatty acid intake from fish," *Epidemiology* 2009, 20:598–603.

136 Hibbeln, JR et al. "Fish consumption and major depression," *Lancet* 1998, https: //pubmed.ncbi.nlm.nih.gov/9643729.

137 Dunstan, JA et al. "Effects of n3 polyunsaturated fatty acid supplementation in pregnancy on maternal and fetal erythrocyte fatty acid composition," *European Journal of Clinical Nutrition* 2004, 58:429–437; Larqué, E. et al. "Docosahexaenoic acid supply in pregnancy affects placental expression of fatty acid transport proteins," *American Journal of Clinical Nutrition* 2006, 84:853–861.

138 Kar, S. et al. "Effects of omega-3 fatty acids in prevention of early preterm delivery: a systematic review and meta-analysis of randomized studies," *European Journal of Obstetrics and Gynecology and Reproductive Biology* 2016, 198:40–46; Muthayya, S. et al. "The effect of fish and omega-3 LCPUFA intake on low birth weight in Indian pregnant women," *European Journal of Clinical Nutrition* 2009, 63:340–346.

139 Burdge GC et al. "Long-chain n3 PUFA in vegetarian women: a metabolic perspective," *Journal of Nutritional Science* 2017, www.ncbi.nlm.nih.gov/pubmed/29209497.

140 Helland, IB et al. "Maternal supplementation with very-long-chain n3 fatty acids during pregnancy and lactation augments children's IQ at 4 years of age," *Pediatrics* 2003, 111:39–44.

141 Kaviani, M. et al. "The Effect of Omega-3 Fatty Acid Supplementation on Maternal Depression during Pregnancy: A Double Blind Randomized Controlled Clinical Trial," *International Journal of Community Based Nursing & Midwifery* 2014, 2:142–147; Makrides, M. et al. "Effect of DHA supplementation during pregnancy on maternal depression and neurodevelopment of young children: a randomized controlled trial," *JAMA* 2010, 304:1675–1683; Su, KP et al. "Omega-3 fatty acids for major depressive disorder during pregnancy: results from a randomized, double-blind, placebo-controlled trial," *Journal of Clinical Psychiatry* 2008, 69:644–651; Nemets, H.

et al. "Omega-3 treatment of childhood depression: A controlled, double-blind pilot study," *Am J Psychiatry* 2006, 163:1098–1100.

142 Monk, C. "Stress and mood disorders during pregnancy: Implications for child development," *Psychiatric Quarterly* 2001, 72:347–357.

143 Lassek, WD, Gaulin, SJC. "Linoleic and docosahexaenoic acids in human milk have opposite relationships with cognitive test performance in a sample of 28 countries," *Prostaglandins Leukot Essent Fatty Acids* 2014; 91:195–201.

144 Lüthi, T. "Nahrung für den Geist," *Neue Zürcher Zeitung*, https://nzzas.nzz.ch/wissen/ernaehrungpisaomega3fettsaeurengehirnld.149010?reduced=true (2.11.2014, zuletzt abgerufen am 31.8.2021).

145 Thuppal, SV et al. "Discrepancy between Knowledge and Perceptions of Dietary Omega-3 Fatty Acid Intake Compared with the Omega-3 Index," *Nutrients* 2017, www.ncbi.nlm.nih.gov/pmc/articles/PMC5622690.

146 Craddock, JC et al. "Algal supplementation of vegetarian eating patterns improves plasma and serum docosahexaenoic acid concentrations and omega-3 indices: a systematic literature review," *Journal of Human Nutrition and Diet* 2017, 30: 693–699.

147 Nehls, M. *Algenöl: Die Ernährungsrevolution aus dem Meer*, Heyne: München, 2018, 158.

148 Avallone, R. et al. "Omega-3 Fatty Acids and Neurodegenerative Diseases: New Evidence in Clinical Trials," *International Journal of Molecular Sciences* 2019, www.ncbi.nlm.nih.gov/pmc/articles/PMC6747747.

149 von Schacky C: "Omega-3 Fatty Acids in Pregnancy: The Case for a Target Omega-3 Index," *Nutrients* 2020, www.ncbi.nlm.nih.gov/pmc/articles/PMC7230742.

150 www.dge.de/ernaehrungspraxis/bevoelkerungsgruppen/schwangere-stillende/handlungsempfehlungen-zur-ernaehrung-in-der-schwangerschaft/?L=0 (last accessed on 20.1.2021).

151 https://eur-lex.europa.eu/legal-content/DE/TXT/?uri=CELEX%3A32017R2470 (last accessed on 20.1.2021).

152 Levenson, CW and Morris, D. "Zinc and neurogenesis: making new neurons from development to adulthood," *Advances in Nutrition* 2011, 2:96–100.

153 Fischer Walker CL et al. "Global and regional child mortality and burden of disease attributable to zinc deficiency," *European Journal of Clinical Nutrition* 2009, 63:591–597.

154 www.zinkorot.de/zinkinlebensmitteln.html.

155 www.dge.de/wissenschaft/referenzwerte/zink.

156 Choudhry, H and Nasrullah, M. "Iodine consumption and cognitive performance: Confirmation of adequate consumption," *Food Science & Nutrition* 2018, 6: 1341–1351.

157 Biban, BG and Lichiardopol, C. "Iodine Deficiency, Still a Global Problem?" *Current Health Sciences Journal* 2017, 43:103–111.

158 Delange, F. "The role of iodine in brain development," *Proceedings of the Nutrition Society* 2000, 59:75–79.

159 Takele, WW et al. "Two-thirds of pregnant women attending antenatal care clinic at the University of Gondar Hospital are found with subclinical iodine deficiency," 2017, https://pubmed.ncbi.nlm.nih.gov/30333053.

160 Plaum P: Immer noch oft Jodmangel in der Schwangerschaft—EUgefördertes Projekt nimmt Politik und Ärzte in die Pflicht. Medscape 2018, https//deutsch.medscape.com/artikelansicht/4907016.

161 MonteroPedrazuela, A. et al. "Modulation of adult hippocampal neurogenesis by thyroid hormones: implications in depressive-like behavior," *Molecular Psychiatry* 2006, 11:361–371.

162 www.dge.de/wissenschaft/referenzwerte/jod.

163 Chen, G. et al. "Enhancement of hippocampal neurogenesis by lithium," *Journal of Neurochemistry* 2000, 75:1729–1734; Fiorentini, A et al. "Lithium improves hippocampal neurogenesis, neuropathology and cognitive functions in APP mutant mice," *PLoS One* 2010, www.ncbi.nlm.nih.gov/pmc/articles/PMC3004858.

164 Marshall TM. "Lithium as a nutrient," *Journal of the American Physicians and Surgeons* 2015, 20:104–109, www.jpands.org/vol20no4/marshall.pdf.

165 Freland, L. and Beaulieu, JM. "Inhibition of GSK3 by lithium, from single molecules to signaling networks," *Frontiers in Molecular Neuroscience* 2012, www.ncbi.nlm.nih. gov/pmc/articles/PMC3282483; Struewing, IT et al. "Lithium increases PGC 1 alpha expression and mitochondrial biogenesis in primary bovine aortic endothelial cells," *FEBS Journal* 2007, 274:2749–2765.

166 Nunes, MA et al. "Microdose lithium treatment stabilized cognitive impairment in patients with Alzheimer's disease," *Current Alzheimer Research* 2013, 10:104–107.

167 Szklarska, D. and Rzymski, P. "Is Lithium a Micronutrient? From Biological Activity and Epidemiological Observation to Food Fortification," *Biological Trace Element Research* 2019, 189:18–27; Takahashi Yanaga, F. "Activator or inhibitor? GSK3 as a new drug target," *Biochemical Pharmacology* 2013, 86:191–199; Wu, Z. and Boss, O. "Targeting PGC1 alpha to control energy homeostasis," *Expert Opinion Therapeutic Targets* 2007, 11:1329–1338.

168 Takahashi Yanaga. "Activator or inhibitor?"; Wu and Boss. "Targeting PGC1 alpha to control energy homeostasis."

169 Kapusta, ND et al. "Lithium in drinking water and suicide mortality," *British Journal of Psychiatry* 2011, 198:346–350; Memon, A. et al. "Association between naturally occurring lithium in drinking water and suicide rates: systematic review and meta-analysis of ecological studies," *British Journal of Psychiatry* 2020, 217:667–678; Spitzer, M. and Graf, H. "Lithium im Trinkwasser—Lithium ins Trinkwasser?," *Geist and Gehirn /Nervenheilkunde* 2010, 3: 157–158; Ohgami, H. et al. "Lithium levels in drinking water and risk of suicide," *British Journal of Psychiatry* 2009, 194:464–465.

170 Mozaffarian, D. "Fish, mercury, selenium and cardiovascular risk: current evidence and unanswered questions," *International Journal of Environmental Research and Public Health* 2009, 6:1894–1916.

171 Conner, TS et al. "Optimal serum selenium concentrations are associated with lower depressive symptoms and negative mood among young adults," *Journal of Nutrition* 2015, 145:59–65; Santos, JR et al. "Nutritional status, oxidative stress and dementia: the role of selenium in Alzheimer's disease," *Frontiers in Aging Neuroscience* 2014, www.ncbi.nlm.nih. gov/pmc/articles/PMC4147716; Varikasuvu, SR et al. "Brain Selenium in Alzheimer's Disease (BRAIN SEAD Study): a Systematic Review and Meta-Analysis," *Biological Trace Element Research* 2019, 189:361–369.

172 Rayman, MP. "The Importance of Selenium to Human Health," *Lancet* 2000, 356:233–241.

173 www.dge.de/wissenschaft/referenzwerte/selen.

174 Beck, MA. "Selenium and host defence toward viruses," *Proceedings of the Nutrition Society* 1999, 58:707–711.

175 Baum, MK et al. "High risk of HIV-related mortality is associated with selenium deficiency," *Journal of Acquired Immune Deficiency Syndromes and Human Retrovirology* 1997, 15:370–374.

176 Moghaddam, A. et al. "Selenium Deficiency Is Associated with Mortality Risk from COVID-19," *Nutrients* 2020, www.ncbi.nlm.nih.gov/pmc/articles/PMC7400921.

177 Nehls, M. "Unified theory of Alzheimer's disease (UTAD): implications for prevention and curative therapy," *Molecular Psychiatry* 2016, www.ncbi.nlm.nih.gov/pmc/articles/PMC4947325.

178 Liu, JL et al. "Iron and Alzheimer's Disease: From Pathogenesis to Therapeutic Implications," *Frontiers in Neuroscience* 2018, www.ncbi.nlm.nih.gov/pmc/articles/PMC6139360.

179 Ayton, S. et al. "Alzheimer's Disease Neuroimaging Initiative. Ferritin levels in the cerebrospinal fluid predict Alzheimer's disease outcomes and are regulated by APOE," *Nature Communications* 2015, www.ncbi.nlm.nih.gov/pmc/articles/PMC4479012.

180 Bredesen, DE et al. "Reversal of cognitive decline in Alzheimer's disease," *Aging* 2016, 8:1250–1258.

181 Stoltzfus, RJ. "Iron deficiency: global prevalence and consequences," *Food Nutrition Bulletin* 2003, 24:99–103.

182 Bastian, TW et al. "Iron Deficiency Impairs Developing Hippocampal Neuron Gene Expression, Energy Metabolism, and Dendrite Complexity," *Developmental Neuroscience* 2016, 38:264–276.

183 www.dge.de/wissenschaft/referenzwerte/eisen.

184 Becker, EW: "Microalgae as a source of protein," *Biotechnology Advances* 2007, 25:207–210.

185 Vazhappilly, R. and Chen, F. "Eicosapentaenoic Acid and Docosahexaenoic Acid Production Potential of Microalgae and Their Heterotrophic Growth," *JAOCS* 1998, 75:393–397; Yongmanitchai, W. and Ward, OP. "Screening of algae for potential

alternative sources of eicosapentaenoic acid," *Phytochemistry* 1991, 9:2963–2967; Subramoniam, A. et al. "Chlorophyll revisited: anti-inflammatory activities of chlorophyll a and inhibition of expression of TNFα gene by the same," *Inflammation* 2012, 35:959–966.

186 Sharma, K. and Schenk, PM. "Rapid induction of omega-3 fatty acids (EPA) in Nannochloropsis sp. by UVC radiation," *Biotechnology and Bioengineering* 2015, 112:1243–1249.

187 Moomaw, W et al. "Cutting Out the Middle Fish: Marine Microalgae as the Next Sustainable Omega-3 Fatty Acids and Protein Source," *Industrial Biotechnology* 2017, 13:234–243.

188 Tessari, P. et al. "Essential amino acids: master regulators of nutrition and environmental footprint?" *Scientific Reports* 2016, www.ncbi.nlm.nih.gov/pmc/articles/PMC4897092.

189 Jones, C. S. et al. "Livestock manure driving stream nitrate," *Ambio* 2019, 48:11431153; Dopelt, K. et al. "Environmental Effects of the Livestock Industry: The Relationship between Knowledge, Attitudes, and Behavior among Students in Israel," *International Journal of Environmental Research and Public Health* 2019, doi: 10.3390/ijerph16081359.

190 Qiu, W. et al. "Folic acid, but not folate, regulates different stages of neurogenesis in the ventral hippocampus of adult female rats," *Journal of Neuroendocrinology* 2019, https://pubmed.ncbi.nlm.nih.gov/31478270; Ferland, G. "Vitamin K and the Nervous System: An Overview of Its Actions," *Advances in Nutrition* 3:204–212.

191 www.vitalstofflexikon.de/VitamineACDEK/VitaminA/Lebe nsmittel.html.

192 Touyarot, K. et al. "A midlife vitamin A supplementation prevents age-related spatial memory deficits and hippocampal neurogenesis alterations through CRABPI," *PLoS One* 2013, https://pubmed.ncbi.nlm.nih.gov/23977218.

193 www.vitalstofflexikon.de/VitaminBKomplex/ThiaminVitaminB1/Lebensmittel.html.

194 Zhao, N. et al. "Impaired hippocampal neurogenesis is involved in cognitive dysfunction induced by thiamine deficiency at early prepathological lesion stage," *Neurobiology of Disease* 2008, 29:176–185.

195 Zeisel, SH and da Costa, KA. "Choline: an essential nutrient for public health," *Nutrition Reviews* 2009, 67:615–623.

196 Norris, J. "Choline in Vegetarian Diets," 2015, http://extension.oregonstate.edu/coos /sites/default/files/FFE/documents/cholinerd.pdf

197 www.vitalstofflexikon.de/VitaminBKomplex/Pantothen saeureVitaminB5/Lebensmittel .html.

198 www.vitalstofflexikon.de/VitaminBKomplex/PyridoxinVitaminB6/Lebensmittel.html.

199 www.vitalstofflexikon.de/VitaminBKomplex/Biotin/Leben smittel.html.

200 Watanabe, F. et al. "Pseudovitamin B(12) is the predominant cobamide of an algal health food, spirulina tablets," *Journal of Agricultural and Food Chemistry* 1999, 47:4736–4741.

201 Wells, ML et al. "Algae as nutritional and functional food sources: revisiting our understanding," *Journal of Applied Phycology* 2017, 29:949–982.

202 Chen, JH and Jiang, SJ. "Determination of cobalamin in nutritive supplements and chlorella foods by capillary electrophoresis inductively coupled plasma mass spectrometry," *Journal of Agricultural and Food Chemistry* 2008, 56:1210–1215.

203 www.vitaminb12.de/tagesbedarf.

204 Shipton, MJ and Thachil, J. "Vitamin B12 deficiency—A 21st century perspective," *Clinical Medicine* 2015, 15:145–150; Guéant, JL et al. "Molecular and cellular effects of vitamin B12 in brain, myocardium and liver through its role as cofactor of methionine synthase," *Biochemie* 2013, 95:1033–1040; Smith, AD. "Hippocampus as a mediator of the role of vitamin B12 in memory," *American Journal of Clinical Nutrition* 2016, 103:959–960.

205 Littlejohns, TJ et al. "Vitamin D and the risk of dementia and Alzheimer disease," *Neurology* 2014, 83:920–928.

206 Durup, D. et al. "A reverse J-shaped association between serum 25-hydroxy vitamin D and cardiovascular disease mortality—the Cop D study," *Journal of Clinical Endocrinology & Metabolism* 2015, www.ncbi.nlm.nih.gov/pubmed/25710567.

207 Ingraham, BA et al. "Molecular basis of the potential of vitamin D to prevent cancer," *Current Medical Research and Opinion* 2008, 24:139–149; Garland, CF et al. "Vitamin D for cancer prevention: global perspective," *Annals of Epidemiology* 2009, 19:468–483.

208 Mons, U. and Brenner, H. "Vitamin D supplementation to the older adult population in Germany has the cost-saving potential of preventing almost 30 000 cancer deaths per year," *Molecular Oncology* 2021, https://pubmed.ncbi.nlm.nih.gov/33540476.

209 Rabenberg, M. et al. "Vitamin D status among adults in Germany—results from the German Health Interview and Examination Survey for Adults (DEGS1)," *BMC Public Health* 2015, www.ncbi.nlm.nih.gov/pmc/articles/PMC4499202.

210 Giovannucci, E. et al. "Prospective study of predictors of vitamin D status and cancer incidence and mortality in men," *Journal of the National Cancer Institute* 2006, 98:451–459.

211 Mohajeri, MH et al. "The role of the microbiome for human health: from basic science to clinical applications," *Eur J Nutr* 2018, 57:1–14.

212 Erny, D. et al. "Host microbiota constantly control maturation and function of microglia in the CNS," *Nature Neuroscience* 2015, www.ncbi.nlm.nih.gov/pubmed/26030851; Liang, S. et al. "Gut-Brain Psychology: Rethinking Psychology From the Microbiota Gut-Brain Axis," *Frontiers in Integrative Neuroscience* 2018, www.ncbi.nlm.nih.gov/pmc/articles/PMC6142822.

213 Ogbonnaya, ES et al. "Adult Hippocampal Neurogenesis Is Regulated by the Microbiome," *Biological Psychiatry* 2015, 78:7–9; Heijtz, RD et al. "Normal gut microbiota modulates brain development and behavior," *PNAS USA* 2011, 108:3047–3052; Desbonnet, L. et al. "Microbiota is essential for social development in the mouse," Mol Psychiatry 2014, 19:146–148.

214 Clemente, JC et al. "The microbiome of uncontacted Amerindians," *Science Advances* 2015, www.ncbi.nlm.nih.gov/pmc/articles/PMC4517851.

215 Gomez, A. et al. "Gut Microbiome of Coexisting BaAka Pygmies and Bantu Reflects Gradients of Traditional Subsistence Patterns," *Cell Reports* 2016, 14:2142–2153.

216 Obregon Tito, A. et al. "Subsistence strategies in traditional societies distinguish gut microbiomes," *Nature Communications* 2015, www.nature.com/articles/ncomms7505.

217 Lingenhöhl, D. "Dein Darm ist, was du isst," www.spektrum.de/news/deindarmist wasduisst/1408962 (4/29/2016, last viewed on 08/31/2021).

218 GBD 2017 Diet Collaborators. "Health effects of dietary risks in 195 countries, 1990–2017: a systematic analysis for the Global Burden of Disease Study 2017," *Lancet* 2019, 393:1958–1972.

219 Bello, MGD et al. "Preserving microbial diversity," *Science* 2018, 362:33–34.

220 Desai, MS et al. "A Dietary Fiber-Deprived Gut Microbiota Degrades the Colonic Mucus Barrier and Enhances Pathogen Susceptibility," *Cell* 2016, 167:1339–1353.

221 Davidson, GL et al. "Diet induces parallel changes to the gut microbiota and problem solving performance in a wild bird," *Scientific Reports* 2020, www.ncbi.nlm.nih.gov /pmc/articles/PMC7699645.

222 Hill, JM et al. "The gastrointestinal tract microbiome and potential link to Alzheimer's disease," *Frontiers in Neurology* 2014, www.ncbi.nlm.nih.gov/pubmed/24772103.

223 Hill, JM and Lukiw, WJ. "Microbial-generated amyloids and Alzheimer's disease (AD)," *Frontiers in Aging Neuroscience* 2015, www.ncbi.nlm.nih.gov/pubmed/25713531.

224 Foster, JA et al. "Gut-brain axis: how the microbiome influences anxiety and depression," *Trends in Neuroscience* 2013, 36:305–312; Valles Colomer, M. et al. "The neuroactive potential of the human gut microbiota in quality of life and depression," *Nature Microbiology* 2019, 4:623–632.

225 Tengeler, AC et al. "Gut microbiota from persons with attention-deficit/hyperactivity disorder affects the brain in mice," *Microbiome* 2020, https://pubmed.ncbi.nlm.nih .gov/32238191.

CHAPTER 12

1 Schaefer, S. et al. "Cognitive performance is improved while walking: Differences in cognitive-sensorimotor couplings between children and young adults," *European Journal of Developmental Psychology* 2010, 7:371–389.

2 Winter, B. et al. "High impact running improves learning," Neurobiol Learn Mem 2007, 87: 597–609; Donnelly, JE et al. "Physical Activity, Fitness, Cognitive Function, and Academic Achievement in Children: A Systematic Review," *Medicine & Science in Sports & Exercise* 2016, 48:1197–1222.

3 Kubesch, S. et al. "A 30-Minute Physical Education Program Improves Students' Executive Attention," *Mind*, Brain and Education 2009, https://onlinelibrary.wiley .com/doi/10.1111/j.1751228X.2009.01076.x.

4 Ellemberg, D. and St Louis Deschênes, M. "The effect of acute physical exercise on cognitive function during development," *Psychology of Sport and Exercise*, 2010, 11:122–126.

5   Tomporowski, PD. "Effects of acute bouts of exercise on cognition," *Acta Psychologica* (Amst) 2003, 112:297–324; Budde, H. et al. "Acute coordinative exercise improves attentional performance in adolescents," *Neuroscience Letters* 2008, 441:219–223; Pesce, C. et al. "Physical activity and mental performance in preadolescents: Effects of acute exercise on free-recall memory," *Mental Health and Physical Activity* 2009, 2:16–22.

6   Hauschild, J. "Schulsport: Bessere Mathenoten dank Ausdauerlauf," www.spektrum .de/news/schulsportbesseremathenotendankaus.

7   Herholz, K et al. "Regional cerebral blood flow in man at rest and during exercise," *Journal of Neurology* 1987, 234, 9–13.

8   Phillips, C. et al. "Neuroprotective effects of physical activity on the brain: a closer look at trophic factor signaling," *Frontiers in Cellular Neuroscience* 2014, www.ncbi .nlm. nih.gov/pubmed/24999318.

9   Fabel, K. et al. "VEGF is necessary for exercise-induced adult hippocampal neurogenesis," *European Journal of Neuroscience* 2003, 18:2803–2812.

10   Black, JE et al. "Learning causes synaptogenesis, whereas motor activity causes angiogenesis, in cerebellar cortex of adult rats," *PNAS USA* 1990, 87:5568–5572.

11   Leiter, O. et al. "The systemic exercise-released chemokine lymphotactin/XCL1 modulates in vitro adult hippocampal precursor cell proliferation and neuronal differentiation," *Scientific Reports* 2019, https://pubmed.ncbi.nlm.nih.gov/31413355.

12   Åberg, D. "Role of the growth hormone/insulin-like growth factor 1 axis in neurogenesis," *Endocrine Developments* 2010, 17:63–76.

13   Yau, SY et al. "Fat cell-secreted adiponectin mediates physical exercise-induced hippocampal neurogenesis: an alternative anti-depressive treatment?," *Neural Regeneration Research* 2015, 10:7–9; Phillips, C. et al. "Neuroprotective effects of physical activity on the brain: a closer look at trophic factor signaling," *Frontiers in Cellular Neuroscience* 2014, www.ncbi.nlm.nih.gov/pubmed/24999318; Katsiki, N. et al. "Adiponectin, lipids and atherosclerosis," Curr Opin Lipidol 2017, 28:347–354.

14   Cao, L. et al. "VEGF links hippocampal activity with neurogenesis, learning and memory," *Nature Genetics* 2004, 36:827–835.

15   John, MJ et al. "Erythropoietin use and abuse," *Indian Journal of Endocrinology and Metabolism* 2012, 16:220–227; Hintz, RL: "Growth hormone: uses and abuses," *BMJ* 2004, 328:907–908.

16   Lin, TW and Kuo, YM. "Exercise benefits brain function: the monoamine connection," *Brain Sciences* 2013, 3:39–53; Gould, E. "Serotonin and hippocampal neuro genesis," *Neuropsychopharmacology* 1999, 21:46–51; Takamura, N. et al. "The effect of dopamine on adult hippocampal neurogenesis," *Progress Neuro-Psychopharmacology and Biological Psychiatry* 2014, 50:116–124

17   Koehl, M. et al. "Exercise-induced promotion of hippocampal cell proliferation requires beta-endorphin," *FASEB Journal* 2008, 22:2253–2262.

18   Griffin ÉW et al. "Aerobic exercise improves hippocampal function and increases BDNF in the serum of young adult males," *Physiology & Behavior* 2011, 104:934941;

Schmolesky MT et al. "The effects of aerobic exercise intensity and duration on levels of brain-derived neurotrophic factor in healthy men," *Journal of Sports Science and Medicine* 2013, 12:502511.

19 Giuffrida, ML et al. "A promising connection between BDNF and Alzheimer's disease," *Aging* (Albany NY) 2018, 10:1791–1792.

20 Smith, JC et al. "Physical activity reduces hippocampal atrophy in elders at gene tic risk for Alzheimer's disease," *Frontiers in Aging Neuroscience* 2014, www.ncbi.nlm.nih. gov/pubmed/24795624.

21 Abbott, RD et al. "Walking and dementia in physically capable elderly men," *JAMA* 2004 Sep 22;292(12):1447–1534.

22 https://openclassrooms.com/en/courses/6663471-develop-your-creativity/6930160-identify-physical-activities-that-nurture-your-creativity (11.3.2020).

23 Chaddock, L. et al. "A neuroimaging investigation of the association between aerobic fitness, hippocampal volume, and memory performance in preadolescent children," *Brain Research* 2010, 1358:172–183; Chaddock, L. et al. "Aerobic fitness and executive control of relational memory in preadolescent children," *Medicine & Science in Sports & Exercise* 2011, 43:344–349.

24 Ghassabian, A. et al. "Determinants of neonatal brain-derived neurotrophic fac tor and association with child development," *Developmental Psychopathology* 2017, 29:1499–1511.

25 Gomes da Silva S et al. "Maternal Exercise during Pregnancy Increases BDNF Levels and Cell Numbers in the Hippocampal Formation but Not in the Cerebral Cortex of Adult Rat Offspring," *PLoS One* 2016, www.ncbi.nlm.nih.gov/pmc/articles/PMC4714851.

26 Molendijk, ML et al. "Serum BDNF concentrations show strong seasonal variation and correlations with the amount of ambient sunlight," *PLoS One* 2012, www.ncbi.nlm.nih.gov/pmc/articles/PMC3487856.

27 Sailani, MR et al. "Lifelong physical activity is associated with promoter hypo methylation of genes involved in metabolism, myogenesis, contractile properties and oxidative stress resistance in aged human skeletal muscle," *Scientific Reports* 2019, www.ncbi.nlm.nih.gov/pmc/articles/PMC6397284.

28 Smith, JA et al. "CaMK activation during exercise is required for histone hyperacetylation and MEF2A binding at the MEF2 site on the Glut4 gene," *American Journal of Physiology Endocrinology and Metabolism* 2008, 295:698–704.

29 Nakajima, K. et al. "Exercise effects on methylation of ASC gene," *International Journal of Sports Medicine* 2010, 31:671–675.

30 Zahl, T. et al. "Physical Activity, Sedentary Behavior, and Symptoms of Major Depression in Middle Childhood," *Pediatrics* 2017, https://pubmed.ncbi.nlm.nih.gov/28069664.

31 Gorham, LS et al. "Involvement in Sports, Hippocampal Volume, and Depressive Symptoms in Children," *Biological Psychiatry: Cognitive Neuroscience and Neuroimaging* 2019, 4:484–492.

32   Erickson KI et al. "Exercise training increases size of hippocampus and improves memory," *PNAS USA* 2011, 108:30173022.

33   Erickson, KI et al. "Physical activity, fitness, and gray matter volume," *Neurobiology Aging* 2014, 35:20–28.

34   Marques, A. et al. "How does academic achievement relate to cardiorespiratory fitness, self-reported physical activity and objectively reported physical activity: a systematic review in children and adolescents aged 6–18 years," *British Journal of Sports Medicine* 2018, https://pubmed.ncbi.nlm.nih.gov/29032365.

35   https://www.aerztezeitung.de/Medizin/Kinderbewegensi-chimmerweni-ger-253800.html (20.3.2019; last accessed on 31.8.2021).

36   Guthold, R. et al. "Worldwide trends in insufficient physical activity from 2001 to 2016: a pooled analysis of 358 population-based surveys with 1·9 million participants," *Lancet Global Health* 2018, 6:1077–1086.

37   Belvederi Murri, M. et al. "Physical Exercise in Major Depression: Reducing the Mortality Gap While Improving Clinical Outcomes," *Frontiers in Psychiatry* 2019, www.ncbi.nlm.nih.gov/pmc/articles/PMC6335323.

38   Dinas, PC et al. "Effects of exercise and physical activity on depression," *Irish Journal of Medical Science* 2011, 180:319–325.

39   http://daten2.verwaltungsportal.de/dateien/seitengenerator/rfb-2013–09-klinikarzt -loellgen-2.pdf.

40   Gilchrist, SC et al. "Association of Sedentary Behavior With Cancer Mortality in Middle-aged and Older US Adults," *JAMA Oncology* 2020, 6:1210 1217.

41   Kyu, HH et al. "Physical activity and risk of breast cancer, colon cancer, diabetes, ischemic heart disease, and ischemic stroke events: systematic review and dose-response meta-analysis for the Global Burden of Disease Study 2013," *BMJ* 2016, www.ncbi .nlm.nih.gov/pmc/articles/PMC4979358.

42   Geissel, W. "Deutsche werden immer mehr zu Bewegungsmuffeln," www.aerztezei tung.de/Panorama/DeutschewerdenimmermehrzuBewegungsmuffeln229370.html (9/5/2018, last viewed on 8/31/2021).

43   www.handelsblatt.com/politik/deutschland/mobilitaet-scheuer-sieht-elektrotretroller-als -echte-alternative-zum-auto/24043912.html (02/27/2019, last viewed on 08/31/2021).

44   Robinson, S. "Experimental studies of physical fitness," *Arbeitsphysiologie* 1938, 10:251–232; Buskirk, ER and Hodgson, JL. "Age and aerobic power: the rate of change in men and women," *Fed Proc* 1987, 46: 1824–1829.

45   Kasch FW et al. "Cardiovascular changes with age and exercise. A 28-year longitudinal study," *Scand J Med Sci Sports* 1995, 5:147151; Kasch FW et al. "Ageing of the cardiovascular system during 33 years of aerobic exercise," *Age Ageing* 1999, 28:531536.

46   Pollock, ML et al. "Effect of age and training on aerobic capacity and body composition of master athletes," *Journal of Applied Physiology* 1987, 62:725–731.

47   McGuire, DK et al. "A 30-year follow-up of the Dallas Bedrest and Training Study: II. Effect of age on cardiovascular adaptation to exercise training," *Circulation* 2001, 104:1358–1366.

48 Shay, JW and Wright, WE. "Hayflick, his limit, and cellular ageing," *Nature Reviews Molecular Cell Biology* 2000, 1:72–76.

49 Oeseburg, H. et al. "Telomere biology in healthy aging and disease," *Pflugers Archiv* 2010, 459:259–268.

50 Arsenis, NC et al. "Physical activity and telomere length: Impact of aging and potential mechanisms of action," *Oncotarget* 2017, 8:45008–45019.

51 Du, M. and Prescott, J. et al. "Physical activity, sedentary behavior, and leukocyte telomere length in women," *American Journal of Epidemiology* 2012, 175:414–422; Brandao, CFC et al. "The effects of short-term combined exercise training on telomere length in obese women: a prospective, interventional study," *Sports Medicine Open* 2020, http://www.ncbi.nlm.nih.gov/pmc/articles/PMC6965549.

CHAPTER 13

1 www.watson.ch/spass/populärkultur/275466438rauchen-macht-gesund-und-glueck lich-sagen-diese-vintage-werbungen (8/10/2016, last viewed on 8/31/2021).

2 Pearce, M. "At least four Marlboro Men have died of smoking-related diseases," *Los Angeles Times*, January 27, 2014, https://www.latimes.com/nation/nationnow/la-na-nn -marlboro-men-20140127-story.html.

3 www.gesundheitstipp.ch/artikel/d/tabakindustrie-hat-raucher-jahrzehntelang -belogen (4/26/2000, last viewed on 8/31/2021).

4 www.who.int/newsroom/factsheets/detail/tobacco (5/27/2020, last viewed on 8/31/2021).

5 Jazbinsek, D. "Tabakkontrolle in Europa: Schlusslicht Deutschland," Dtsch Arztebl 2017, 114: A837/B707/C693

6 www.aerztezeitung.de/Medizin/Kinderaerztefor dernRau chverbotim Auto300717 .html (9/9/2017, last viewed on 08/31/2021).

7 www.who.int/newsroom/factsheets/detail/tobacco (05/27/2020).

8 Streeck, H. et al. "Infection fatality rate of SARS-CoV-2 in a superspreading event in Germany," *Nature Communications* 2020, https://www.nature.com/articles /s41467020 19509y; Ioannidis, JPA. "Infection fatality rate of COVID19 inferred from sero-prevalence data," *Bulletin of the World Health Organization* 2021, 99:19–33.

9 Wang, A. et al. "Active and passive smoking in relation to lung cancer incidence in the Women's Health Initiative Observational Study prospective cohort," *Annals of Oncology* 2015, 26:221–230.

10 www.who.int/newsroom/factsheets/detail/tobacco (5/27/2020, last viewed on 8/31/2021).

11 Oberg, M. et al. "Worldwide burden of disease from exposure to secondhand smoke: a retrospective analysis of data from 192 countries," *Lancet* 2011, 377:139–146.

12 https://ourworldindata.org/smoking (June 2021, last viewed on 09/16/2021).

13 Batty, GD et al. "Effect of maternal smoking during pregnancy on offspring's cognitive ability: empirical evidence for complete confounding in the US National Longitudinal Survey of Youth," *Pediatrics* 2006, 118:943–950.

14   Liao, Y. et al. "Bilateral frontoparietal integrity in young chronic cigarette smokers: a diffusion tensor imaging study," *PLoS One* 2011, www.ncbi.nlm.nih.gov/pubmed /22069452.

15   Galván, A. et al. "Neural correlates of response inhibition and cigarette smoking in late adolescence," *Neuropsychopharmacology* 2011, 36:970–978.

16   Yolton, K. et al. "Exposure to environmental tobacco smoke and cognitive abilities of U.S. children and adolescents," *Environ Health Perspect* 2005, 113:98–103; Park, S. et al. "Environmental tobacco smoke exposure and children's intelligence at 811 years of age," *Environmental Health Perspectives* 2014, 122:1123–1128.

17   Bélanger, M. et al. "Nicotine dependence symptoms among young never smokers exposed to secondhand tobacco smoke," *Addict Behav* 2008, 33:1557–1563.

18   www.uclahealth.org/news/tobacco-smoking-impacts-teens-brains-ucla-study-shows (03/2/2011, last viewed on 8/31/2021).

19   Durazzo, TC et al. "Cigarette smoking is associated with cortical thinning in anterior frontal regions, insula and regions showing atrophy in early Alzheimer's Disease," *Drug and Alcohol Dependence* 2018, 192:277–284.

20   Abrous DN et al. "Nicotine self-administration impairs hippocampal plasticity," *Journal of Neuroscience* 2002, 22:3656–3662.

21   Durazzo TC et al. "Interactive effects of chronic cigarette smoking and age on hippocampal volumes," *Drug and Alcohol Dependence* 2013, 133:704711; Hawkins KA et al. "The effect of age and smoking on the hippocampus and memory in late middle age," *Hippocampus* 2018, 28:846–849.

22   Health Effects Institute, "State of Global Air 2019," https://www.stateofglobalair.org /sites/default/files/soga_2019_report.pdf.

23   www.who.int/news/item/29–10-2018-more-than-90-of-the-world's-childrenbreathe -toxic-air-every-day.

24   www.who.int/ceh/publications/air-pollution-child-health/en (29.10.2018, last view on 31.08.2021).

25   https://deutsch.medscape.com/artikelansicht/4907410 (11/5/2018, last viewed on 08/31/2021).

26   D'Angiulli, A. "Severe urban outdoor air pollution and children's structural and functional brain development, from evidence to precautionary strategic action," *Frontiers in Public Health* 2018, www.ncbi.nlm.nih.gov/pubmed/29670873.

27   de Prado Bert, P. et al. "The Effects of Air Pollution on the Brain: A Review of Studies Interfacing Environmental Epidemiology and Neuroimaging," *Current Environmental Health Reports* 2018, 5:351–364.

28   Oudin, A. et al. "Association between neighbourhood air pollution concentrations and dispensed medication for psychiatric disorders in a large longitudinal cohort of Swedish children and adolescents," *BMJ Open* 2016, www.ncbi.nlm. nih.gov/pmc /articles/PMC4893847.

29   Suades González, E. et al. "Air Pollution and Neuropsychological Development: A Review of the Latest Evidence," *Endocrinology* 2015, 156:3473–3482.

30   Morishita, M. et al. "Effect of Portable Air Filtration Systems on Personal Exposure to Fine Particulate Matter and Blood Pressure Among Residents in a Low Income Senior Facility: A Randomized Clinical Trial," *JAMA Internal Medicine* 2018, 178:1350–1357.

31   Calderón Garcidueñas, L. et al. "Long-term air pollution exposure is associated with neuroinflammation, an altered innate immune response, disruption of the blood-brain barrier, ultrafine particulate deposition, and accumulation of amyloid beta-42 and alpha-synuclein in children and young adults," *Toxicologic Pathology* 2008, 36:289–310; Younan, D. et al. "Particulate matter and episodic memory decline mediated by early neuroanatomic biomarkers of Alzheimer's disease," *Brain* 2020, 143:289–302.

32   Gale, SD et al. "Association between exposure to air pollution and prefrontal cortical volume in adults: A cross-sectional study from the UK Biobank," *Environmental Research* 2020, https://pubmed.ncbi.nlm.nih.gov/32222630.

33   Hedges, DW et al. "Association between exposure to air pollution and hippocampal volume in adults in the UK Biobank," *Neurotoxicology* 2019, 74:108–120.

34   Möykkynen, T. and Korpi, ER. "Acute effects of ethanol on glutamate receptors," *Basic & Clinical Pharmacology & Toxicology* 2012, 111:4–13.

35   Taffe, MA et al. "Long-lasting reduction in hippocampal neurogenesis by alcohol consumption in adolescent nonhuman primates," *PNAS USA* 2010, 107:11104–11109.

36   Morris, SA et al. "Alcohol inhibition of neurogenesis: a mechanism of hippocampal neurodegeneration in an adolescent alcohol abuse model," *Hippocampus* 2010, 20:596–607.

37   Meda, SA et al. "Longitudinal Effects of Alcohol Consumption on the Hippocampus and Parahippocampus in College Students," *Biological Psychiatry: Cognitive Neuroscience and Neuroimaging* 2018, 3:610–617.

38   Solfrizzi, V et al. "Alcohol consumption, mild cognitive impairment, and progression to dementia," *Neurology* 2007, 68:1790–1799.

39   Pfefferbaum, A et al. "Frontal lobe volume loss observed with magnetic resonance imaging in older chronic alcoholics," *Alcohol, Clinical and Experimental Research* 1997, 21:521–529.

40   Yücel, M. et al. "Hippocampal harms, protection and recovery following regular cannabis use," *Translational Psychiatry* 2016, www.ncbi.nlm.nih.gov/pmc/articles/PMC 5068875.

41   Albaugh, MD et al. "Association of Cannabis Use During Adolescence With Neurodevelopment," *JAMA Psychiatry* 2021, www.ncbi.nlm.nih.gov/pmc/artic les/PMC 8209561.

42   https://deutsch.medscape.com/artikelansicht/4910110 (6/28/2021, last viewed on 08/31/2021).

43   Jiang, W. et al. "Cannabinoids promote embryonic and adult hippocampus neurogenesis and produce anxiolytic and antidepressant-like effects," *Journal of Clinical Investigation* 2005, 115:3104–3116.

44 Silote, GP et al. "Emerging evidence for the antidepressant effect of cannabidiol and the underlying molecular mechanisms," *Journal of Chemical Neuroanatomy* 2019, 98:104–116; Watt, G., Karl, T. "In vivo Evidence for Therapeutic Properties of Cannabidiol (CBD) for Alzheimer's Disease," *Frontiers in Pharmacology* 2017, www.ncbi.nlm.nih. gov/pmc/articles/PMC5289988.

45 Swift, W. et al. "Analysis of cannabis seizures in NSW, Australia: cannabis potency and cannabinoid profile," *PLoS One* 2013, www.ncbi.nlm.nih.gov/pmc/articles/PMC 3722200.

46 Dos Santos, A. et al.: "Methylmercury and brain development: A review of recent literature," *Journal of Trace Elements in Medicine and Biology* 2016, 38:99–107.

47 Rossetti, MF et al. "Agrochemicals and neurogenesis," *Molecular and Cellular Endocrinology* 2020, https://pubmed.ncbi.nlm.nih.gov/32315720.

48 www.sciencemediacenter.de/alleangebote/rapidreaction/details/news/vorderentsche idungzurglyphosatzulassung (11/7/2017, last viewed on 08/31/2021).

49 Mills, PJ et al. "Excretion of the Herbicide Glyphosate in Older Adults Between 1993 and 2016," *JAMA* 2017, 318:1610–1611.

50 Bensch, K. "Der eine Satz von Minister Schmidt," www.tagesschau.de/inland/glyphosat -zulassung-schmidt-101.html (11/28/2017, last viewed on 08/31/2021).

51 www.badenwuerttemberg.de/de/service/presse/pressemitteilung/pid/ergebnisse-zum -oekomonitoring-2012-vorgestellt (06/24/2013, last accessed on 8/31/2021); https: //oekomonitoring.uabw.de/berichte.html.

52 Wilhelm, M. et al. "Occurrence of perfluorinated compounds (PFCs) in drinking water of North Rhine Westphalia, Germany and new approach to assess drinking water contamination by shorter-chained C4C7 PFCs," *International Journal of Hygiene and Environmental Health* 2010, 213:224–232; Foguth, R. et al. "Per and Polyfluoroalkyl Substances (PFAS) Neurotoxicity in Sentinel and Non-Traditional Laboratory Model Systems: Potential Utility in Predicting Adverse Outcomes in Human Health," *Toxics* 2020, www.ncbi.nlm.nih.gov/pmc/articles/PMC7355795.

53 Kougias, DG et al. "Perinatal exposure to an environmentally relevant mixture of phthalates results in a lower number of neurons and synapses in the medial prefrontal cortex and decreased cognitive flexibility in adult male and female rats," *Journal of Neuroscience* 2018, http://www.jneurosci.org/content/38/31/6864.

54 Li, N. et al. "Identifying periods of susceptibility to the impact of phthalates on children's cognitive abilities," *Environmental Research* 2019, 172:604–614; Li, N. et al. "Gestational and childhood exposure to phthalates and child behavior," Environ Int 2020, https://pubmed.ncbi.nlm.nih.gov/32798801.

55 Dönges, J.: "Babyflaschen setzen große Mengen an Mikroplastik frei," www.spektrum .de/news/babyf laschen-setzen-grosse-mengen-an-mikroplastikfrei/1783613 (19.10.2020, zuletzt abgerufen am 31.08.2021); www.aerztezeitung. de/Panorama/Bor-im-Spielschleim -Phthalate-in-Kunststoffpuppen-411056.html (7/7/2020, last viewed on 08/31/2021).

56   Reddam, A. and Volz, DC. "Inhalation of two Prop 65-listed chemicals within vehicles may be associated with increased cancer risk," *Environ Int* 2021, https://pubmed.ncbi.nlm.nih.gov/33524670.

57   Kim, MS et al. "Organic solvent metabolite, 1,2 diacetylbenzene, impairs neural progenitor cells and hippocampal neurogenesis," *Chemical-Biological Interactions* 2011, 194:139–147.

58   Thrasher, JD and Kilburn, KH. "Embryo toxicity and teratogenicity of formaldehyde," *Archives of Environmental Health* 2001, 56:300–311.

59   Mohapel, P. et al. "Forebrain acetylcholine regulates adult hippocampal neurogenesis and learning," *Neurobiology of Aging* 2005, 26:939–946.

60   Pfistermeister, B. et al. "Anticholinergic burden and cognitive function in a large German cohort of hospitalized geriatric patients," *PLoS One* 2017, www.ncbi.nlm.nih.gov/pmc/articles/PMC5302450.

61   www.aerzteblatt.de/nachrichten/61606/StudiesiehtAnticholinergikaalsDemenzrisiko (1/27/2015, last viewed on 08/31/2021).

62   www.pharmazeutische-zeitung.de/ausgabe-412013/erkennen-erklaeren-ersetzen (2.10.2013, last accessed on 31.08.2021).

63   www.nih.gov/newsevents/newsreleases/nihfundedstudysuggestsacetaminophenexposurepregnancylinkedhigherriskadhdautism (10/30/2019, last viewed on 8/31/2021).

64   www.nih.gov/news-events/news-releases/nih-funded-study-suggests-acetaminophen-exposure-pregnancy-linked-higher-risk-adhd-autism (30.10.2019, last accessed on 31.8.2021).

65   Chantong, C et al. "Increases of proinflammatory cytokine expression in hippocampus following chronic paracetamol treatment in rats," *Asian Archive of Pathology* 2013, www.asianarchpath.com/view/48.

66   Mischkowski, D. et al. "A Social Analgesic? Acetaminophen (Paracetamol) Reduces Positive Empathy," *Frontiers in Psychology* 2019, www.ncbi.nlm.nih.gov/pmc/articles/PMC6455058.

67   Hutton, JS et al. "Associations Between Screen-Based Media Use and Brain White Matter Integrity in Preschool-Aged Children," *JAMA Pediatrics* 2020, www.ncbi.nlm.nih.gov/pmc/articles/PMC6830442.

68   https://cmhd.northwestern.edu/wpcontent/uploads/2015/06/ParentingAgeDigitalTechnology.REVISED.FINAL_.2014.pdf (2014); www.commonsensemedia.org/sites/default/files/uploads/research/commonsenseparentcensus_executivesummary_forweb.pdf (2016).

69   Misra, S. et al. "The iPhone effect: the quality of in-person social interactions in the presence of mobile devices," Environ Behav 2016, 48:275–298; McDaniel, Brandon T. et al. "'Technoference' and Implications for Mothers' and Fathers' Couple and Coparenting Relationship Quality," *Computers in Human Behavior* Vol. 80 (2018): 303–313; Hutton, JS et al. "Associations Between Screen-Based Media Use and Brain White Matter Integrity in Preschool-Aged Children," *JAMA Pediatrics* 2020, www.ncbi.nlm.nih.gov/pmc/articles/PMC6830442.

70   Przybylski, AK and Weinstein, N. "Can you connect with me now? How the presence of mobile communication technology influences face-to-face conversation quality," *Journal of Social and Personal Relationships* 2013, 30:237–246.

71   McDaniel, BT and Radesky, JS. "Technoference: longitudinal associations between parent technology use, parenting stress, and child behavior problems," *Pediatric Research* 2018, 84:210–218.

72   www.diagnosemedia.org/artikel/detail&id=6 (6/25/2018, last viewed on 8/31/2021).

73   Spitzer, M. "www (WeltWeite Werbung[1]) und die Folgen Radikalisierung, Spi onage, Vertrauens und Wahrheitsverlust," *Nervenheilkunde* 2018, 37:303–311.

74   Shoukat, S. "Cell phone addiction and psychological and physiological health in adolescents," *EXCLI Journal* 2019, 18:47–50; Alhassan, AA et al. "The relationship between addiction to smartphone usage and depression among adults: a cross sectional study," *BMC Psychiatry* 2018, www.ncbi.nlm.nih.gov/pmc/articles/PMC5970452.

75   www.diagnosemedia.org/artikel/detail&id=6 (6/25/2018, last viewed on 8/31/2021).

76   Bühring, P.: "Internetabhängigkeit: Dem realen Leben entschwunden," www.aerzteblatt .de/archiv/184492/Internetabhaengigkeit-Dem-realen-Leben-entschwunden (2016, last viewed on 08/31/2021).

77   Schmid, R.: "Jedes Jahr erkranken 20.000 Kinder," www.aerztezeitung.de/Politik /Jedes-Jahre-erkranken-20000-Kinder-305725.html (8.3.2017, last accessed on 31. 08.2021).

78   Subramanian, KR. "Product Promotion in an Era of Shrinking Attention Span," *International Journal of Engineering and Management Research* 2017, 7:8591.

79   http://mitbestimmungzh.ch/data/documents/181102SpiegelD ieserMistverdirbtun-salle.pdf (3.11.2018).

80   Mugunthan, N. et al. "Effects of long-term exposure of 9001800 MHz radiation emitted from 2G mobile phone on mice hippocampus—a histomorphometric study," *Journal of Clinical and Diagnostic Research* 2016, 10:1–6.

81   McClelland, S. III and Jaboin, JJ. "The Radiation Safety of 5G WiFi: Reassuring or Russian Roulette?" *International Journal of Radiation Oncology, Biology, Physics* 2018, 101:1274–1275.

82   Moon, JH. "Health effects of electromagnetic fields on children," *Clinical and Experimental Pediatrics* 2020, 63:422–428.

83   www.diagnosefunk.org /publikationen/artikel/detail& newsid=1739 (8/13/2021, last viewed on 08/31/2021).

CHAPTER 14

1   Libet, B. et al. "Time of conscious intention to act in relation to onset of cerebral activity (readiness-potential). The unconscious initiation of a freely voluntary act," *Brain* 1983, 106:623–642.

2   https://news.mongabay.com/2022/11/deforestation-is-pushing-amazon-to-point-of-no -return-wwf-report/; Manisalidis, I. et al. "Environmental and Health Impacts of Air Pollution: A Review," *Frontiers in Public Health* 2020, doi: 10.3389/fpubh .2020.00014.

3   Bratsberg, B. and Rogeberg, O. "Flynn effect and its reversal are both environmentally caused," *PNAS USA* 2018, J115:6674–6678.

4   Koch, C. and Crick, F. "The zombie within," *Nature* 2001, 411:893.

5   https://plato.stanford.edu/entries/zombies ((3/19/2019, last viewed on 8/31/2021).

6   Koch and Crick. "The zombie within."

7   Solms, M. *The Hidden Spring: A Journey to the Source of Consciousness*, Profile Books: London, 2021.

8   Ayan, S. "Der Autopilot im Kopf," www.spektrum.de/news/wieentstehtbewusstsein /15891 4 6 (05/28/2021, last viewed at 8/31/2021)

9   Davis, MH. "Measuring individual differences in empathy: Evidence for a multidimensional approach," *Journal of Personality and Social Psychology* 1983, 44:113–126.

10   Konrath, SH et al. "Changes in Dispositional Empathy in American College Students Over Time: A Meta-Analysis," *Personality and Social Psychology Review* 2011, 15:180–198.

11   Ceballos, G. et al. "Biological annihilation via the ongoing sixth mass extinction signaled by vertebrate population losses and declines," *PNAS USA* 2017, 114:6089–6096.

12   Jordan, B. "Scientists Say Humans' 'Lack of Empathy' Is Leading to Global Species Annihilation," Alternet, December 18, 2017, https://www.alternet.org/2017/12 /scientists-say-humans-lack-empathy-leading-global-species-annihilation.

13   Twenge, JM and Foster, JD. "Birth Cohort Increases in Narcissistic Personality Traits Among American College Students, 1982–2009," *Social Psychological and Personality Science* 2010, 1: 99–106.

14   Hartmann, C. "Wir, wir, wir!," www.spektrum.de/news/kollektiver-narziss-mus-kann -demokratien-gefaehrden/1659300 (7/15/2019, last viewed on 08/31/2021).

15   Thompson, D. "Google's CEO: 'The Laws Are Written by Lobbyists,'" www.theatlantic .com/technology/archive/2010/10/googles-ceo-thelaws-are-written-by-lobbyists /63908 (1.10.2010, last accessed on 31.8.2021).

16   www.hohenloheungefiltert.de/?p=26759 (6/8/2020, last viewed on 08/31/2021).

17   www.smartcitydialog.de/wpcontent/uploads/2020/03/LangfassungSmartCity Charta2017.pdf.

18   Ebd, S. 42

19   Ebd., S. 43

20   Kramper, G. "Totale Überwachung—China will Noten an alle seine Bürger verteilen," www.stern.de/digital/technik/chinatotaleueberwachungsosollenallebuergerbewertetwerden 7943770.html (4/17/2018, last viewed on 08/31/2021).

21   www.amnesty.de/informieren/aktuell/chinaderdressiertemens ch (7/25/2019).

22   von Lüpke, M. and Harms, F. "Im schlimmsten Fall kollabiert unsere Weltordnung," www.tonline.de/nachrichten/wissen/geschichte/id_88582030/hararizurpandemi ecoronahatdaspotentialdieweltbesserzumachen.html (10/23/2020, last viewed on 8/31/2021).

23   Schwab, K. and Malleret, T. *COVID-19: Der Große Umbruch*, Forum Publishing: Köln/Genf 2020.

24   Knobbe, M. "Vertrauliche Regierungsstudie beschreibt CoronaSzenarien für Deutschland,"www.spiegel.de/politik/deutschland/coronaindeutschlandvertraulicheregier ungsstudiebeschreibtverschiedenesze n arienalcafaacl 3932434db4de2f63bce0315d (3/27/2020, last viewed at 8/31/2021).

25   Chomsky, N. *Media Control*, Nomen Verlag: Frankfurt am Main 2021 (Klappentexte)

26   Wolfangel, E. "Das richtige Gefühl," www.spektrum.de/news/emotionenper fektionierenkuenstlicheintelligenz/1566366 (5/23/2018, last viewed at the 8/31/2021); Krauß, P. and Maier, A. "Der Geist in der Maschine," www.spektrum. de/magazin /bewusstekidergeistindermaschine/1875787 (16/16/2021, last viewed on 08/31/2021)

27   www.theguardian.com/technology/2014/feb/22/computersclevererthan humans15years (22.2.2014, zuletzt abgerufen am 31.8.2021).

28   www.openinsights.de/kuenstlicheintelligenzwennderfortschri ttploetzlich explodiert (1/13/2018).

29   www.independent.co.uk/news/science/stephenhawkingtr anscendencelooks implicationsartificialintelligencearewetakingaiseriouslye nough9313474. html (10/23/2017, last viewed on 8/31/2021).

30   Nehls, M. *Die MethusalemStrategie: Vermeiden, was uns daran hindert, gesund älter und weiser zu werden*, Mental Enterprises: Vörstetten 2011

31   Nehls, M. *Die MethusalemStrategie: Vermeiden, was uns daran hindert, gesund älter und weiser zu werden*, Mental Enterprises: Vörstetten 2011

32   Nehls, M. *Die AlzheimerLüge: Die Wahrheit über eine vermeidbare Krankheit*, Heyne: München 2014

33   Nehls, M. *Das CoronaSyndrom: Wie das Virus unsere Schwächen offenlegt—und wie wir uns nachhaltig schützen können*, Heyne: München 2021

34   Abbott, A. "COVID's mental health toll: how scientists are tracking a surge in depression," *Nature* 2021, 590:194–195; Pérez, S. et al. "Levels and variables associated with psychological distress during confinement due to the coronavirus pandemic in a community sample of Spanish adults," *Clinical Psychology & Pyschotherapry* 2021, 28:606–614; Soldevila Domenech, N. et al. "Effects of COVID19 Home Confinement on Mental Health in Individuals with Increased Risk of Alzheimer's Disease," *Journal of Alzheimer's Disease* 2021, 79:1015–1021.

35   Riess, H. "The Science of Empathy," *J Patient Exp* 2017, 4:74–77.

IMAGE CREDITS

Figure 1—page 4: Phineas Gage, Picture from Hey Paul Studios. License: CC BY 2.0.

Figure 2—page 4: Cerebrum, https://www.istockphoto.com/fr/vectoriel/lobes-des-r%C3%A9gions-du-cerveau-humain-%C3%A9tiquet%C3%A9s-illustration-gm2161123529-581590512.

Figure 3—page 22: Hippocampus, https://de.wikipedia.org/wiki/Hippocampus#/media/Datei:Hippocampus_and_seahorse_cropped.JPG.

Figure 4—page 22: Henry Molaison, Rochelle Schwartz-Bloom, Duke University; https://sites.duke.edu/apep/module-3-alcohol-cell-suicide-and-the-adolescent-brain/explore-more/the-incredible-story-of-h-m/.

Figure 5—page 26: Walter Freeman performing lobotomy, source: Getty Images.

Figure 6—page 33: Image created by author.

Figure 7—page 34: Image created by author.

Figure 8—page 58: Image created by author.

Figure 9—page 59: Image based on data from the article "Developmental Trajectories of Amygdala and Hippocampus from Infancy to Early Adulthood in Healthy Individuals" by Uemtasu, Akiko, et al. *PLos One*, 2012. https://pmc.ncbi.nlm.nih.gov/articles/PMC3467280/.

Figure 10—page 61: Image based on data from the article "Changes in Dispositional Empathy in American College Students Over Time: A Meta-Analysis" by Konrath, Sara, et al. *Personality and Social Psychology Review*, 2011, 15: 180–198. https://www.academia.edu/5053162/Changes_In_Dispositional_Empathy_Sara_Konrath.

Figure 11—page 75: Image based on data from the article "Fatty Acid Composition of Human Brain Phospholipids during Normal Development" by Martinze, M., et al. *Journal of Neurochemistry*, 1998. https://pubmed.ncbi.nlm.nih.gov/9832152/.

Figure 12—page 77: Image based on data from the study "Higher RBC EPA + DHA Corresponds with Larger Total Brain and Hippocampal Volumes" by Pottala, James V., et al. *Neurology*, 2014. https://pmc.ncbi.nlm.nih.gov/articles/PMC3917688/.

Figure 13—page 79: Battle of the frontal lobes, source: Shutterstock.

Figure 14—page 82: Vitruvius, image source: Adobe Stock.

Figure 15—page 95: Image created by author.

Figure 16—page 111: Competing child schemes, source: iStock.

Figure 17—page 119: Image based on data from the article "Sleep and Sleep Disordered Breathing in Hospitalized Patients" by Knauert, Melissa P. et al., *Seminars in Respiratory and Critical Care Medicine*, 2014. https://pubmed.ncbi.nlm.nih.gov/25353103/.

Figure 18—page 124: Information adapted from the article "Epidemiology of the Human Circadian Clock" by Roenneber, T. et al., *Sleep Medicine Review*, 2007. https://pubmed.ncbi.nlm.nih.gov/17936039/.

Figure 19—page 125: Information adapted from the article "Epidemiology of the Human Circadian Clock" by Roenneber, T. et al., *Sleep Medicine Review*, 2007. https://pubmed.ncbi.nlm.nih.gov/17936039/.

Figure 20—page 138: Image created by author.

Figure 21—page 147: Image created by author.

Figure 22—page 148: Torture over milk, source: Shutterstock.

Figure 23—page 154: Image created by author.

Figure 24—page 156: Image adapted from Nehls, Michael, *The Algae Oil Revolution: Fight Disease and Promote Brain Development and Mental Health with the Vegan Elixir From the Sea*. Skyhorse 2025.

Figure 25—page 193: Images created by author.

Figure 26—page 194: Image created by author.

Figure 27—page 195: No smarter than a yeast fungus?, source: Shutterstock.

INDEX